Emerging Trends in Mechatronics

Series Editor

Aydin Azizi, Oxford, UK

Ajay Kumar · Parveen Kumar · Aydin Azizi
Editors

Advances in Mechatronics Systems

Principles, Elements and Applications

Springer

Editors
Ajay Kumar
Department of Mechanical Engineering
JECRC University
Jaipur, Rajasthan, India

Parveen Kumar
Department of Mechanical Engineering
Rawal Institute of Engineering
and Technology
Faridabad, Haryana, India

Aydin Azizi
School of Engineering, Computing
and Mathematics
Oxford Brookes University
Oxford, UK

ISSN 2731-4855 ISSN 2731-4863 (electronic)
Emerging Trends in Mechatronics
ISBN 978-981-97-8471-4 ISBN 978-981-97-8472-1 (eBook)
https://doi.org/10.1007/978-981-97-8472-1

This Springer imprint is published by the registered company Springer Nature Singapore Pte Ltd.
The registered company address is: 152 Beach Road, #21-01/04 Gateway East, Singapore 189721, Singapore

Preface

Mechatronics is the synergetic integration of mechanical engineering with electrical engineering and/or electronics and possibly with other disciplines for the purposes of design, manufacture, operation, and maintenance of a product. It is mainly composed of mechanical engineering; electronic engineering, electrical, computer science, information engineering, and intelligent technology. The term "Mechatronics" first appeared in the supplement of the Japanese magazine mechanical design in 1971. With the rapid development and wide application of computer technology, mechatronics technology has achieved unprecedented development. Now mechatronics technology is machinery Microelectronics and information are the products of the integration and intersection of these technologies. The main objective of this interdisciplinary engineering field is the study of automata from an engineering perspective, thinking about the design of products and manufacturing processes and systems. Mechatronics systems are well established and executed within a great number of industries including aircraft, automotive, and aerospace industries; machine tools, computers, electronics, semiconductors and communications, and biomedical, aviation, military, civil, agriculture, etc. Thus mechatronics is a response to the industry's growing need for engineers who can work across the boundaries of certain engineering disciplines to find and employ the best possible technology combinations for the best possible solutions to today's increasingly difficult engineering problems.

This book is an edited compilation of original research and applications backed with experimental, numerical and theoretical data on mechatronics systems, contributed by researchers, academics, scientists, industrialists and professionals from the field on various aspects of mechatronics. Therefore, it will provide logical, technical as well as analytical solutions and ideas to complex problems faced by researchers in the field of mechatronics. The academicians and students will get a comprehensive update on the state-of-art in this area and ample ideas for further research and innovation in mechatronics. This book has the following key features:

- Presents numerical, experimental and computational approach for signal sensing, fault detection and diagnosis.
- Presents different concepts from the cross-disciplinary field of mechatronics, including discussions from mechanical engineering, electrical engineering, electronics engineering, and computer science.
- Includes original theoretical, experimental, and modeling results of advancements in mechatronics systems.

The book consists of 11 chapters that describe perspectives of Advances in Mechatronics Systems.

- Chapter "Design of an Iterative Model for Fault Severity Classification Using CNN-LSTM and Attention Mechanisms" explores how the design model integrates CNNs for feature extraction, LSTMs for sequence learning, and attention mechanisms to iteratively improve fault severity classification accuracy, enhancing system reliability and predictive maintenance.
- Chapter "Event-Triggered Optimal Output Tracking Control for Industrial Automation with Stochastic Dynamics" deals with design of optimal controller for industrial automation subject to stochastic uncertain dynamics.
- Chapter "Wind Energy Converters and Controllers: Practical Aspects" explores innovative control strategies and advanced technologies for wind energy systems, focusing on enhancing efficiency, power quality, and grid stability amidst variable wind conditions.
- Chapter "Dual-Band Planar Antenna Design Using Characteristics Mode Analysis for 5G Sub-6 Band and WLAN Application" discusses how the Dual-Band Planar Antenna demonstrates desirable attributes for dual-band operation, making it well-suited for 5G and WLAN applications.
- Chapter "Automation in Friction Stir Welding: A New Era of Fabrication Techniques" explores the advancements, applications, and challenges of robotic friction stir welding, and highlighting its versatility and precision in fabricating technology.
- Chapter "Comparative Analysis of OFDM and FBMC System Using Cognitive Radio Technique" focuses on integrating the energy detection schemes with OFDM and FBMC waveforms and compares the simulations and comparative analysis of FBMC and OFDM with different aspects to analyze spectrum sensing.
- Chapter "Integration of Convolutional Neural Networks for Autonomous Guidance in Automated Guided Vehicles" emphasizes the significance of sophisticated machine learning methods in industrial automation and establishes a basis for future research and development in this domain.
- Chapter "Survey on Drones and IOT Devices for Futuristic Smart City" aims to demonstrate how cooperative drone technology and the Internet of Things (IoT) could improve the intelligence of smart cities.
- Chapter "Contingency Assessment in Power System Having Wind Energy Resource Integration with IEEE 14 Bus System" presents a detailed contingency assessment of the IEEE 14 bus system integrated with wind energy, analyzing

system stability and reliability using performance indices to identify critical scenarios and enhance grid operations.
- Chapter "Mathematical Modelling and Speed Control of SRM Drive" presents the mathematical modelling and small signal modelling of switched reluctance machine.
- Chapter "Robotic Arm Inverse Kinematics Through Graphical Approach" deals with the forward and inverse kinematics of a four degree of freedom spatial robotic arm.

The editors acknowledge the professional support received from Springer and express their gratitude for this opportunity.

Reader's observations, suggestions, and queries are welcome.

Jaipur, India — Ajay Kumar
Faridabad, India — Parveen Kumar
Oxford, UK — Aydin Azizi

Acknowledgements

The editors are grateful to Springer for showing their interest in publishing this book in the buzz area of Advances in Mechatronics Systems. The editors express their personal adulation and gratitude to Loyola D'Silva (Publishing Editor) Springer, for giving consent to publish our work. He undoubtedly imparted great and adept experience in terms of systematic and methodical staff who have helped the editors to compile and finalize the manuscript. The editors also extend their gratitude to Ms. Vinothini Elango, (Project Coordinator—Total Service, Books Production), Springer, for support during her tenure.

The editors, wish to thank all the chapter authors for contributing their valuable research and experience to compile this volume. The chapter authors, corresponding author in particular, deserve special acknowledgments for bearing with the editors, who persistently kept bothering them with deadlines, and with their remarks.

Dr. Ajay also wishes to express his gratitude to his parents, Sh. Jagdish and Smt. Kamla, and his loving brother Sh. Parveen for their true and endless support. They have made him able to walk tall before the world regardless of sacrificing their happiness and living in a small village. He cannot close these prefatory remarks without expressing his deep sense of gratitude and reverence to his life partner Mrs. Sarita Rathee for her understanding, care, support, and encouragement to keep his moral high all the time. No magnitude of words can ever quantify the love and gratitude he feels in thanking his daughters, Sejal Rathee and Mahi Rathee, and son Kushal Rathee who are the world's best children.

Finally, the editors obligate this work to the divine creator and express their indebtedness to the "almighty" for gifting them the power to yield their ideas and concepts into substantial manifestation. The editors believe that this book would enlighten the readers about each feature and characteristic of Advances in Mechatronics Systems-Principles, Elements, and Applications.

Ajay Kumar
Parveen Kumar
Aydin Azizi

Contents

Editors and Contributors

About the Editors

Ajay Kumar is currently serving as a Professor in Department of Mechanical Engineering, School of Engineering and Technology, JECRC University, Jaipur, Rajasthan, India. He received his Ph.D. in the field of Advanced Manufacturing from Guru Jambheshwar University of Science and Technology, Hisar, India after B.Tech. (Hons.) in mechanical engineering and M.Tech. (Distinction) in manufacturing and automation. His areas of research include Incremental Sheet Forming, Artificial Intelligence, Sustainable Materials, Additive Manufacturing, Mechatronics, Smart Manufacturing, Industry 4.0, Waste Management, and Optimization Techniques. He has over 120 publications in international journals of repute including SCOPUS, Web of Science, and SCI indexed database and refereed international conferences. He has also co-authored and co-edited many books with ELSEVIER, CRC Press, WILEY, De Gruyter, and IGI Global and conference proceedings with IOP. He has organized various national and international events including an International conference on Mechatronics and Artificial Intelligence (ICMAI-2021), International conference on Artificial Intelligence, Advanced Materials, and Mechatronics Systems (AIAMMS-2023) as conference chair. He has more than 20 national and international patents to his credit. He has supervised more than eight M.Tech., Ph.D. scholars and numerous undergraduate projects/thesis. He has a total of 15 years of experience in teaching and research. He is a Guest Editor and Review Editor of reputed journals including Frontiers in Sustainability. He has contributed to many international conferences/symposiums as a session chair, expert speaker, and member of the editorial board. He has won several proficiency awards during the course of his career, including merit awards, best teacher awards, and so on. He is an adviser of QCFI, Delhi Chapter student cell at JECRC University, and has also authored many in-house course notes, lab manuals, monographs, and invited chapters in books. He has organized a series of Faculty Development Programs, International Conferences, workshops, and seminars for researchers, Ph.D., UG and PG level students. He is associated with many research, academic, and professional societies in various capacities.

https://scholar.google.co.in/citations?user=TmZS4JIAAAAJ&hl=en
https://www.webofscience.com/wos/author/record/D-5813-2019
Publons Profile: https://publons.com/researcher/1596469/ajay-kumar/
Research Gate: https://www.researchgate.net/profile/Ajay_Kumar349
Researcher ID: D-5813-2019
http://www.researcherid.com/rid/D-5813-2019

Parveen Kumar is currently serving as an Assistant Professor and Head in Department of Mechanical Engineering at Rawal Institute of Engineering and Technology, Faridabad, Haryana, India. He has submitted Ph.D. from National Institute of Technology, Kurukshetra, Haryana, India. He completed his B.Tech. (Hons.) from Kurukshetra University, Kurukshetra, India and M.Tech. (Distinction) in Manufacturing and Automation from Maharshi Dayanand University, Rohtak, India. His areas of research include Intelligent Manufacturing Systems, Materials; Die less Forming, Design of Automotive systems, Additive Manufacturing, CAD/CAM and Artificial Intelligence, machine learning and Internet of Things in manufacturing, multi objective optimization techniques. He has over 50 publications in international journals of repute including SCOPUS, Web of Science, and SCI indexed database and refereed international conferences. He has eight national and international patents in his credit. He has supervised two M.Tech. scholars and numerous undergraduate projects/theses. He has a total of 13 years of experience in teaching and research. He has co-authored/co-edited several books with ELSEVIER, CRC Press, WILEY, De Gruyter, and IGI Global. He is also a book series editor of "Industrial and Manufacturing Systems and Technologies: Sustainable and Intelligent Perspectives" in CRC Press, Taylor and Francis. He has organized a series of Faculty Development Programmes, workshops, and seminar for researcher, UG level students. https://scholar.google.com/citations?hl=en&authuser=1&user=ESQ-RnYAAAAJ

Aydin Azizi holds a Ph.D. degree in Mechanical Engineering. Certified as an official instructor for the Siemens Mechatronic Certification Program (SMSCP), he currently serves as a Senior Lecturer at Oxford Brookes University. His current research focuses on investigating and developing novel techniques to model, control, and optimize complex systems. Dr. Azizi's areas of expertise include Control and Automation, Artificial Intelligence, and Simulation Techniques. Dr. Azizi is the recipient of the National Research Award of Oman for his AI-focused research, DELL EMC's "Envision the Future" completion award in IoT for "Automated Irrigation System", and 'Exceptional Talent' recognition by the British Royal Academy of Engineering.
https://scholar.google.com/citations?user=CoQco90AAAAJ&hl=en

Contributors

Nikhil Agrawal Indian Institute of Information Technology, Nagpur, Nagpur, Maharashtra, India

Sourabh Anand University School of Information, Communication and Technology, Guru Gobind Singh Indraprastha University, New Delhi, Dwarka, India

Nancy Arya Department of CSE, SGT University, Gurgaon, Haryana, India

R. Ashutosh DigiTele Networks Pvt. Ltd, Hyderabad, Telangana, India

Payal Bansal Poornima University, Jaipur, India

Pushpendra S. Bharti University School of Information, Communication and Technology, Guru Gobind Singh Indraprastha University, New Delhi, Dwarka, India

J. Kumar Chandra DigiTele Networks Pvt. Ltd, Hyderabad, Telangana, India

Vinita Chellappan Government Engineering College Palakkad, APJ Abdul Kalam Technological University, Palakkad, Kerala, India

Tausif Diwan Indian Institute of Information Technology, Nagpur, Nagpur, Maharashtra, India

Jayprakash Giri Electrical Engineering, NIMS University, Jaipur, Rajasthan, India

Sunil Gupta School of Computer and Systems Sciences, Jaipur National University, Jaipur, India

Mohammad Farukh Hashmi National Institute of Technology, Warangal, Warangal, Telangana, India

Satnam Kaur Department of CSE, SGT University, Gurgaon, Haryana, India

Jyoti C. Kolte Poornima University, Jaipur, India

Ajay Kumar Department of Mechanical Engineering, School of Engineering and Technology, JECRC University, Jaipur, Rajasthan, India

Ashwini Kumar Poornima University, Jaipur, India

Mukul Kumar University School of Information, Communication and Technology, Guru Gobind Singh Indraprastha University, New Delhi, Dwarka, India

Parveen Kumar Department of Mechanical Engineering, Rawal Institute of Engineering and Technology, Faridabad, Haryana, India

Renu Kumawat Computer and Communication Engineering, Manipal University Jaipur, Jaipur, India

Stella Kurian Government Engineering College Palakkad, APJ Abdul Kalam Technological University, Palakkad, Kerala, India

G. Madan DigiTele Networks Pvt. Ltd, Hyderabad, Telangana, India

Neeraj Kumar Mishra Electrical Engineering, NIMS University, Jaipur, Rajasthan, India

Mayur Parate Indian Institute of Information Technology, Nagpur, Nagpur, Maharashtra, India

Y. V. S. V. Puneeth DigiTele Networks Pvt. Ltd, Hyderabad, Telangana, India

Asit Baran Puri Department of Mechanical Engineering, National Institute of Technology Durgapur, Durgapur, West Bengal, India

Gajendra Singh Rajawat Department of Information Technology, Poornima College of Engineering, Jaipur, India

Shashi Rathore Indian Institute of Information Technology, Nagpur, Nagpur, Maharashtra, India;
Shri Ramdeobaba College of Engineering and Management, Nagpur, Maharashtra, India

Parul Sahare Indian Institute of Information Technology, Nagpur, Nagpur, Maharashtra, India

Virendra Swaroop Sangtani Department of Electrical Engineering, Swami Keshvanand Institute of Technology, Management and Gramothan, Jagatpura, Jaipur, India

Manoj Kumar Satyarthi University School of Information, Communication and Technology, Guru Gobind Singh Indraprastha University, New Delhi, Dwarka, India

Meghnath Sen Department of Mechanical Engineering, National Institute of Technology Durgapur, Durgapur, West Bengal, India

Dinesh Sethi Electronics and Communication Department, JECRC University, Jaipur, India

Girraj Sharma Electronics and Communication Department, JECRC University, Jaipur, India

Brijraj Singh Solanki Department of Computer Engineering, Poornima College of Engineering, Jaipur, India

Rupayali Swaroop Electronics and Communication Department, JECRC University, Jaipur, India

Ashwani Tapde TIEIT, Bhopal, Madhya Pradesh, India

A. Venkata Vishnu DigiTele Networks Pvt. Ltd, Hyderabad, Telangana, India

Design of an Iterative Model for Fault Severity Classification Using CNN-LSTM and Attention Mechanisms

Shashi Rathore, Parul Sahare, Mayur Parate, Nikhil Agrawal, Tausif Diwan, and Mohammad Farukh Hashmi

Abstract Accurate classification of ball bearing fault severity is imperative for predictive maintenance and mechanical system reliability. A new hybrid deep learning methodology at the junction of deep convolutional neural networks and long short-term memory recurrent neural networks, empowered by attention mechanisms and domain-specific feature engineering, is proposed to further enhance fault severity classification accuracy. The proposed approach will address the limitations of the available methodologies, which were oriented either to the spatial or temporal analysis of vibration signals. Spatial patterns are captured by the CNN, while at the same time, temporal dependencies are dealt with by the LSTM. Further emphasis on only the fault-related important features is brought out by an attention layer. Additionally, domain-specific feature engineering is performed, wherein statistical and frequency domain features obtained using the Fast Fourier Transform aid in increasing the accuracy of classification. The model works with an accuracy of up to 94%, outperforming both standalone CNN and LSTM models. Finally, explainable AI methods—more specifically, SHAP values and LIME—improve the interpretability of the results from the model by explaining how features contribute to

S. Rathore (✉) · P. Sahare · M. Parate · N. Agrawal · T. Diwan
Indian Institute of Information Technology, Nagpur, Nagpur, Maharashtra, India
e-mail: shashirathore@gmail.com; dtea22ece001@iiitn.ac.in

P. Sahare
e-mail: parulsahare@iiitn.ac.in

M. Parate
e-mail: mparate@iiitn.ac.in

N. Agrawal
e-mail: nagrawal@iiitn.ac.in

T. Diwan
e-mail: tdiwan@iiitn.ac.in

S. Rathore
Shri Ramdeobaba College of Engineering and Management, Nagpur, Maharashtra, India

M. F. Hashmi
National Institute of Technology, Warangal, Warangal, Telangana, India
e-mail: mdfarukh@nitw.ac.in

A. Kumar et al. (eds.), *Advances in Mechatronics Systems*, Emerging Trends in Mechatronics, https://doi.org/10.1007/978-981-97-8472-1_1

predictions and confirming the central role of temporal patterns. This methodology advances fault severity classification with pragmatic implications for predictive maintenance and mechanical system reliability improvement through the combined power of CNN, LSTM, attention mechanisms, and domain-specific feature engineering processes.

Keywords Fault severity · Predictive maintenance · Hybrid deep learning · Convolutional neural networks · Long short-term memory · Attention mechanisms

1 Introduction

The reliability and effectiveness of mechanical systems in any industrial application are very important because an unscheduled breakdown results in huge losses in terms of time and finances. Ball bearings are one of the most important components in these systems, influencing their smooth operation. However, they are also prone to various failures, which may evolve from minor into severe if not detected early enough for rectification. Accurate identification of the fault severity in ball bearings is an important task in predictive maintenance so that timely intervention will avoid catastrophic failure and offer extended life to the machinery [1–3]. Traditional approaches towards fault diagnosis focused on the signal processing arena and conventional machine learning models. These traditional methods extract either statistical features or frequency domain characteristics from vibration signals, usually with classifiers based on SVM or decision trees. Currently, they have limited success, for the simple reason that these complicated vibration signal patterns cannot be effectively captured. For this, both spatial features, which are greatly informative about the structure of the signal, and temporal dependencies, which give an understanding of the dynamics of faults, are very important for accurate classification. Traditional techniques poorly capture these aspects.

Deep learning methods, especially Convolutional Neural Networks and Long Short-Term Memory networks, have shown great promise in countering these limitations. One of the inherent strengths of CNNs is in spatial feature extraction from raw input data, impounding patterns, and structures of various fault types. Evidence has been reported on the strengths of LSTMs in capturing temporal dependencies and can, therefore, be widely applied in time series analyses where the order and timing of events are crucial. Although a CNN or an LSTM can be trained individually, their results cannot capture the complementary information of spatial and temporal dimensions underlying vibration signals. With that in mind, this chapter presents a brand-new hybrid model that integrates a CNN effectively with an LSTM for fault severity classification. First, the preprocessed vibration signal is fed into the hybrid CNN-LSTM model to extract the spatial features through the CNN layers. These features are then fed into LSTM layers, which capture the temporal dynamics of the signals. This combination will then allow the model to learn more robust representations of the input data, hence leading to better classification performance. These initial experiments using this hybrid model resulted in quite a big accuracy increase from 85 to 92% compared to using only either CNN or LSTM for different cases [4–6] and scenarios.

An attention mechanism is integrated into this CNN-LSTM architecture in order to enhance even better the performance of this model. Attention mechanisms have become very popular in a lot of domains due to their functionality, which involves dynamic concentration on the most relevant parts of the input data in order to ensure effectiveness in a model. Properly, an attention layer weighs features that have been extracted by the convolutional neural network and the long short-term memory in accordance with their relevance for classification. It enables the model to focus on critical fault-related patterns, improving the accuracy to 94%.

Complementing all the deep learning elements, domain-specific feature engineering is also used to bring more insights into the model. Statistical features derived from vibration signals in the form of mean, standard deviation, skewness, and kurtosis give a very fine representation of the signal characteristics. Besides, frequency domain features obtained by the Fast Fourier Transform capture periodicities and harmonics indicative of faults. Integrating these into the created hybrid model shows better accuracy in classification, which is the case in fusing deep learning with domain knowledge.

Then Explainable AI techniques are applied so as to have the model interpretable and easily adaptable into practical settings. SHAP values and LIME are applied to explain how much each feature contributes to the model's predictions. SHAP values give global explanations about the importance of features in general, while LIME explains local instances of predictions. The methods confirm the really important roles that some frequency domain features and temporal patterns take in fault classification. This result will offer useful insight into how model validation and refinement must be performed.

It provides a holistic framework of fault severity classification in ball bearings, considering the integration of CNN-LSTM hybrid models with attention mechanisms, domain-specific feature engineering, and explainable AI techniques. This iterative design of the model, other than improving the state of the art in fault diagnosis, also brings practical improvements into predictive maintenance strategies by enhancing the reliability and efficiency of mechanical systems.

Model Design and Contribution

This chapter gives a model design that combines convolutional neural networks with long short-term memory networks, further improved by an attention mechanism and domain-specific feature engineering in handling fault severity classification tasks in ball bearings. In other words, it combines the capability of the CNN for spatial feature extraction with the temporal sequence learning capabilities of LSTMs and the dynamic focusing power of attention mechanisms, while statistical and frequency-domain features are effectively grasped for more comprehensive analysis.

The model's first module is responsible for preprocessing data on the vibration signal in order to enhance the quality and relevance of the input features. This preprocessing involves noise reduction, normalization, and segmentation for the raw vibration signals. The segregated signals are then fed into the layers of a CNN, which convolve the same to bring out spatial features. Spatial features show signal patterns related to different fault conditions. The CNN layers can be designed to hold multiple filters of varying sizes to capture a diverse set of spatial characteristics. These are followed by pooling layers, bringing down the dimensionality and computational complexity, and retaining only the most salient features for further processing.

These obtained spatial features enter the LSTM layers after the layers of CNN. Among the many choices available, LSTM neural networks are particularly chosen for their inherent ability to handle temporal dependencies. Such dependencies are the most important aspects in understanding fault progression and its worsening with time. Due to their gating mechanism, which allows for memorization of the previous states, these enable picking up long-term dependencies within signal sequences. This becomes even more important when fault severity classification is based on the temporal evolution of vibration patterns, which conveys vital information about the condition of ball bearings.

Further to enhance the performance of the model, an attention mechanism is integrated into the CNN-LSTM architecture. The attention layer dynamically computes the weights for features depending on their relevance to any given classification task. The attention mechanism is very instrumental in making the model pay attention to the most relevant part of the input data, hence giving priority to critical fault-related patterns that improve the accuracy of the classification. No subtle yet important viewpoints, which may foreshadow the incidence or development of faults, will be ignored.

Besides deep learning components, domain-specific feature engineering can also be leveraged to add more insights into the model. It extracts statistical features, which include mean, standard deviation, skewness, and kurtosis, from vibration signals. All these features that are extracted in this section will give a clear understanding of the characteristics of these vibration signals, which have a very good capability of capturing the overall behavior of the signal, thus leading to different fault conditions. The frequency domain features are obtained from time-domain signals through Fast Fourier Transform. This transformation will allow the model to pick up periodicities and harmonics often related to specific fault types. Integrating these into input hybridization strengthens model differentiation between normal, minor fault, and severe fault conditions.

In the course of the implementation, alladin Execute XAI techniques to guarantee model interpretability, which ensures its application in practical scenarios. SHAP values present the global view of feature importance for how each feature contributes to the prediction of the model. This will establish the most critical features in fault severity classification and validate the decision-making process of the model. Moreover, LIME takes care of local explanations for the model in individual predictions. LIME perturbs the input data and trains an interpretable model on these perturbations to approximate the behavior of the deep learning model around particular instances. This approach thus adds insight into the model's behavior on a case-by-case basis, confirming that temporal patterns captured by the LSTM are important in distinguishing between minor and severe faults.

Its contribution lies in how it is able to exercise manifold advanced techniques in order to realize high accuracy in fault severity classification. This involves the integration of not only deep architectures of CNN and LSTM in capturing complex patterns within the vibration signals but also the attention mechanisms that will make the model focus more on those patterns bearing the most important information. The attention mechanism serves to further strengthen the focus on key features of the model, while on the other hand, XAI techniques clearly explain how model predictions are made. To that effect, it has advanced the state of the art in fault diagnosis and also offers practical benefits when applying predictive maintenance strategies. Attention mechanisms simply put this iterative model designed to go a long way in improving fault severity classification. We provide a model that combines the benefits of CNNs and LSTMs with an attention mechanism and domain-specific feature engineering, which not only encapsulates spatial and temporal features of the vibration signal but also leads to better accuracy in classification. Moreover, XAI makes sure that the predictions of this model are interpretable and trustworthy so they can be used without any doubt in practical applications. This work demonstrates that hybrid deep models can become a promising means of improving the reliability and efficiency of mechanical systems, and the development of predictive maintenance strategies.

2 Review of Existing Models for Fault Severity Classification of Ball Bearings

In the last few decades, a good number of techniques have been developed for the fault diagnosis and classification of ball bearings; some pertinent literature is reviewed here to bring to the forefront some of the methodologies that influenced the design of the proposed attention-based CNN-LSTM hybrid model. Ke et al. [1] proposed the usage of a genetic algorithm-based adaptive suppression method in the approach against CEEMD modal mixing; it can optimize the decomposition process for motor bearing fault diagnosis. The approach enhances the clarity of the features of faults within the vibration signals through the minimization of mode mixing. This work demonstrates that genetic algorithms are appropriate for the optimization of signal processing techniques and hence create a platform for integrating different machine learning algorithms with the methods of empirical mode decomposition.

Xia et al. [2] presented another approach for the diagnosis of spindle bearing faults using an improved underdamped asymmetric bistable stochastic resonance. The method presented by uses underdamped stochastic resonance, which is able to heighten feeble fault signals to a detectable magnitude. The method's effectiveness is illustrated by using it for vibration signal analysis, especially in the time-domain zero-crossing index. The current chapter has therefore underpinned the role of innovative signal processing approaches to fault detection capability enhancement. Balani and Chavan [3] presented a method for early fault diagnosis based on the improved kurtogram with Gaussian Mixture Model (GMM). The approach combines spectral kurtosis with GMM for feature extraction and classification to provide accuracy in fault detection at the early stages. Principal component analysis refines the features extracted, reducing the dimensionality and improving the performance of the model. The current study therefore reveals the potential of integrating statistical feature extraction with advanced machine learning classifiers.

Hamadache et al. [7] investigated rotor speed-based bearing fault diagnosis with varying speed and constant load conditions. Their method uses PCA with sum square error for fault detection and proves to be more robust in dynamic operating conditions. This approach underlines the development of diagnostic techniques for adaption in a practical operating environment. Wang et al. [8] applied wavelet packet denoising combined with random forests for fault diagnosis of rolling bearings. Wavelet packet denoising reduces noise to a larger degree and the random forests classifier deals with complicated relationships among features. The approach is therefore enough to be practical in the industry since it achieves very high accuracy even in very noisy situations. This paper argues that noise reduction and robust algorithms of classification are necessary for fault diagnosis.

Cui et al. [9] proposed a fault severity classification and size estimation method using the vibration mechanism of a ball bearing. Only then, with nonlinear dynamic modeling, is this approach likely to give detailed fault characteristics and their progression. Ultimately, it is only when the understanding of fault dynamics is enhanced that it is likely to aid in much more accurate classification and prognostics. Kuncan [4] proposed a bearing fault diagnosis approach combining one-dimensional local binary patterns with grey relational analysis. This intelligent technique exploits the ease of 1D-LBP in feature extraction and the efficiency of GRA in classifying them to ensure high diagnosis accuracy. The results showed that the hybrid model has bright prospects in fault detection.

Ojaghi et al. [5] presented an improved analytical model of induction motors with localized bearing faults, where the modified winding function approach has been developed. The developed model simulates the impacts of faulted conditions, including the influence of localized faults on motor performance. A simulation-based approach in fault diagnosis proves the significance associated with theoretical models in understanding and hence predicting fault behavior. Abdelkader et al. [6] presented an improved rolling bearing fault diagnosis using a more effective denoising method based on complete ensemble empirical mode decomposition with optimized thresholding operation. Since this approach enhances the capability of extracting fault features from noisy signals, advanced denoising techniques can hence considerably improve diagnostic accuracy.

Yan et al. [10] proposed a feature selection framework-based multiscale morphological analysis algorithm for fault diagnosis of rolling bearings. The approach integrates multiscale morphological filtering with grey correlation analysis for feature selection and improves the performance in classification. In this approach, the authors focus on feature selection to improve the model performance. Jian et al. [11] applied chaotic dynamic errors in bearing fault diagnosis by fusing convolutional neural networks with short-time Fourier transform. This approach embroils the chaos characteristics of the vibration signals; hence, it realizes fault detection with high accuracy. The integration of CNNs with time–frequency analysis techniques underlines the potential of deep learning in fault diagnosis. Yan et al. [12] proposed a new scheme for chaotic feature extraction by smooth multi-scale morphological filtering and composite multi-scale morphological fractal dimension for early fault diagnosis of rolling bearings. This method provides better performance in detecting faults at an incipient stage by catching the fractal characteristics of the vibration signals. One then has the question of complex signal patterns as significant in the application of fractal analysis in fault diagnosis. Cui et al. [13] proposed a quantitative and localization diagnosis method for defective ball bearings using vertical-horizontal synchronizing signal analysis. In this way, one will nail down the location and severity of faults with details and give the diagnostic models more accuracy. It shows the primary significance of spatial relationships in fault detection using synchronization signal analysis. He et al. [14] developed a plastic-bearing fault diagnosis method using two-step data mining. The method fuses the techniques of time–frequency analysis with pattern recognition to achieve high accuracy during fault detection. Combining data mining techniques with signal processing methods reveals that an integrated approach can substantially accomplish fault diagnosis.

Mahesh et al. [15] proposed a data-driven intelligent condition adaptation method using deep responsible active learning for bearing fault detection. In this method, deep convolutional neural networks have been used along with active learning to achieve higher accuracy in fault detection. Active learning shows the manifold potential of adaptive models in real-time fault diagnosis. Raj and Murali [16] proposed an early classification method for bearing faults using morphological operators and fuzzy inference. This realizes the early detection of faults through a combination of signal processing with fuzzy logic techniques. In fault diagnosis, integration by fuzzy systems underscores the issue of handling uncertainty in diagnostic models. Liao et al. [17] proposed an attention-embedded quadratic network, called Qttention, for effective and interpretable bearing fault diagnosis. In this approach, quadratic convolutional neural network with attention mechanisms is integrated to improve both accuracy and interpretability. Attention mechanisms contribute specifically to focusing on relevant features in otherwise complex models.

Dubey et al. [18] developed an automated variational nonlinear chirp mode decomposition method for bearing fault diagnosis. The method gives a better way of capturing nonlinear characteristics of the vibration signal and hence gives high diagnostic accuracy. The application of Chirp mode decomposition underlines the need for capturing changing signal variations in fault diagnosis. Zheng et al. [19] proposed a deep-domain generalization method incorporating a priori diagnosis knowledge for cross-domain fault diagnosis of rolling bearings. This approach enhances the generalization ability of the models in domains other than where they are trained, which originates the potential of transfer learning in fault diagnosis. Nazari et al. [20] developed a current noise cancellation method based on the integration of stator-synchronized currents for bearing fault diagnosis. Consequently, these methods reduce noise and increase the clarity of fault features. The resort to noise cancellation techniques underlines the importance of signal clarity in diagnostic models (Table 1).

Harmouche et al. [21] improved the ball bearings fault diagnosis by employing the global spectrum of vibration signals. Coupling envelope analysis with linear discriminant analysis, the approach realized a high diagnostic accuracy. The application of the global spectrum analysis reflects that comprehensive analysis of signals is very important in fault diagnosis. Frosini et al. [22] proposed an induction machine-bearing fault detection method based on statistical processing of stray flux measurements. This method captures the effect of bearing faults on stray flux, thereby accomplishing high diagnosis accuracy. The use of a statistical processing technique in this method proves the importance of feature extraction in fault diagnosis. Jin et al. [23] developed a motor-bearing fault diagnosis method using trace ratio linear discriminant analysis. It is obvious from the methodology developed in this chapter that it has a better feature selection and classification accuracy, thus showing the prospect of linear discriminant analysis in fault diagnosis. Kim and Chai [24] proposed a bearing fault diagnosis method based on common-domain data and support vector machines. This approach improves traditional cross-domain fault diagnosis and illustrates the potential for adapting domain techniques. Kim and Kim [25] proposed a physics-constrained time–frequency fusion network that combines attention mechanisms to ensure noise-robust bearing fault diagnosis, which fuses attribution methods of CNNs with time–frequency analysis for improved accuracy and robustness. The use of attention mechanisms elaborates the need for concentration on relevant features in noisy environments.

All of these studies, taken together, underline the progress in fault diagnosis methodologies from signal processing techniques, to machine learning algorithms through deep learning models, to explainable AI methods. In this context, an attention-based CNN-LSTM hybrid model is proposed wherein the strengths of each methodology shall be combined to come up with high accuracy, robustness, and interpretability in fault severity classification for ball bearings. Concretely, the comprehensive design of the model and its thorough evaluation prove it to have highly likely practical applications in predictive maintenance, thereby much contributing to the field of fault diagnosis.

3 Proposed Design of an Iterative Model for Fault Severity Classification Using CNN-LSTM and Attention Mechanisms

In view of the prevalent problems of high complexity and low efficiency of fault detection, the existing methods can be designed with an iterative model of fault severity classification using CNN-LSTM with attention mechanisms. As shown in Fig. 1, at

Table 1 Empirical review of existing methods

Reference number	Research objectives	Outcomes	Methodology utilized in context of fault severity classification of ball bearings
[1]	Optimize mode mixing in CEEMD using genetic algorithms for motor bearing fault diagnosis	Improved accuracy in fault diagnosis by reducing mode mixing	Adaptive suppression method using genetic algorithm in Complete Ensemble Empirical Mode Decomposition (CEEMD)
[2]	Amplify weak fault signals in spindle bearings using underdamped stochastic resonance	Enhanced detection of weak fault signals	Underdamped asymmetric bistable stochastic resonance method
[3]	Detect early-stage faults using enhanced kurtogram and GMM	High accuracy in early-stage fault detection	Enhanced kurtogram and Gaussian Mixture Model (GMM) with Principal Component Analysis (PCA)
[7]	Diagnose faults under variable speed and constant load conditions	Robust fault detection under varying operational conditions	Rotor speed-based bearing fault diagnosis (RSB-BFD) with Principal Component Analysis (PCA) and sum square error
[8]	Achieve high accuracy under noisy conditions using wavelet packet denoising	High accuracy in noisy environments	Wavelet packet denoising combined with random forests classifier
[9]	Classify fault severity and estimate fault size using vibration mechanism	Detailed insights into fault characteristics and progression	Nonlinear dynamic modeling based on vibration mechanism
[4]	Diagnose bearing faults using 1D-LBP and GRA	High diagnostic accuracy	One-dimensional local binary pattern (1D-LBP) combined with grey relational analysis (GRA)
[5]	Simulate effects of localized faults in induction motors	Accurate simulation of localized fault impacts	Analytic model using modified winding function theory
[6]	Enhance fault feature extraction from noisy signals	Improved diagnostic accuracy	Denoising method based on complete ensemble empirical mode decomposition with optimized thresholding operation

(continued)

Table 1 (continued)

Reference number	Research objectives	Outcomes	Methodology utilized in context of fault severity classification of ball bearings
[10]	Enhance feature selection and classification using multiscale morphological analysis	Improved model performance	Feature selection framework-based multiscale morphological analysis algorithm with grey correlation analysis
[11]	Capture chaotic nature of vibration signals for fault detection	High accuracy in fault detection	Chaotic dynamic errors for bearing fault diagnosis integrating CNNs with short-time Fourier transform (STFT)
[12]	Detect early-stage faults using fractal characteristics of vibration signals	Enhanced detection of early-stage faults	Chaotic feature extraction with smooth multi-scale morphological filtering (SMMF) and composite multi-scale morphological fractal dimension (CMMFD)
[13]	Provide quantitative and localization diagnosis of defective ball bearings	Detailed insights into fault location and severity	Vertical-horizontal synchronization signal analysis
[14]	Diagnose plastic bearing faults using time–frequency analysis and pattern recognition	High accuracy in fault detection	Two-step data mining approach integrating time–frequency analysis with pattern recognition techniques
[15]	Improve fault detection accuracy using deep responsible active learning	Enhanced fault detection accuracy	Data-driven intelligent condition adaptation method using CNNs and active learning
[16]	Achieve early fault detection using morphological operators and fuzzy inference	Early fault detection achieved	Morphological operators combined with fuzzy inference system
[17]	Enhance accuracy and interpretability in bearing fault diagnosis	High accuracy and interpretability	Attention-embedded quadratic network (Qttention) integrating quadratic CNNs with attention mechanisms
[18]	Capture nonlinear characteristics of vibration signals for fault diagnosis	High diagnostic accuracy	Automated variational nonlinear chirp mode decomposition method

(continued)

Table 1 (continued)

Reference number	Research objectives	Outcomes	Methodology utilized in context of fault severity classification of ball bearings
[19]	Enhance model generalization across different domains	Improved cross-domain fault diagnosis	Deep domain generalization method combining a priori diagnosis knowledge
[20]	Reduce noise for clearer fault feature extraction	Enhanced clarity of fault features	Current noise cancellation method integrating stator synchronized currents
[21]	Integrate envelope analysis and LDA for fault diagnosis	High diagnostic accuracy	Fault diagnosis method based on global spectrum of vibration signals integrating envelope analysis with linear discriminant analysis (LDA)
[22]	Detect bearing faults using statistical processing of stray flux measurements	High diagnostic accuracy	Induction machine bearing fault detection using statistical processing of stray flux measurements
[23]	Enhance feature selection and classification accuracy	Improved classification accuracy	Motor bearing fault diagnosis using trace ratio linear discriminant analysis
[24]	Improve cross-domain fault diagnosis using common-domain data and support vector machines	Enhanced cross-domain fault diagnosis	Fault diagnosis method using common-domain data and support vector machines
[25]	Enhance accuracy and robustness in noisy environments	High accuracy and robustness	Physics-informed time–frequency fusion network with attention mechanisms
[26–36]	Machine learning and ANN in diversified field	Prediction of accurate results through machine learning and deep learning methods	CNN, Random-Forest algorithm, Logistic-Regression algorithm, SVM, KNN, Long Short-Term Memory (LSTMS), Recurrent Neural Network (LSTM-RNN), Multilayer Perceptron (MLP) and Decision-Tree algorithm, etc.

the very beginning, it integrates a frequency domain feature extraction process that is rather critical to the vibration signal in fault severity classification for ball bearings. The method transforms time-domain signals into frequency domains and tries to capture the essential periodicities and harmonics indicative of various fault conditions. First, preprocessing is done on the raw vibration signal for noise reduction and normalization to ensure that the signal being passed is clean for analysis. Fast Fourier Transform is the core transformation technique used in efficiently transforming the time-domain signals into the frequency-domain representation. The fundamental process by which a time-domain signal, is transformed into the frequency domain is defined by the Fourier Transform using Eq. 1:

$$X(f) = \int_{-\infty}^{\infty} x(t)e^{-j2\pi ft}dt \tag{1}$$

This operation represents the continuous Fourier Transform, where $X(f)$ is the frequency domain representation of the signal, $x(t)$ is the time-domain signal, f is the frequency, and j is the imaginary unit. The Fourier Transform decomposes the signal into its constituent frequencies, providing a comprehensive view of its spectral components. However, in practical applications, the Discrete Fourier Transform (DFT) is used, as the signals are sampled and finite. The DFT of a discrete signal $x[n]$ with N samples is expressed via Eq. 2,

$$X[k] = \sum_{n=0}^{N-1} x[n]e^{-\frac{j2\pi}{N}kn} \tag{2}$$

where, $X[k]$ is the frequency domain representation, $x[n]$ is the discrete time-domain signal, N is the number of samples, and k is the index representing the discrete frequencies for the signals. The FFT algorithm efficiently computes the DFT, reducing the computational complexity from $O(N^2)$ to $O(N\log N)$ making it suitable for real-time applications.

Once the signal is transformed into the frequency domain, various features are extracted to characterize the signal's frequency content. One key feature is the Power Spectral Density (PSD), which quantifies the power present at different frequencies. The PSD is calculated via Eq. 3,

$$\text{PSD}(f) = \frac{|X(f)|^2}{T} \tag{3}$$

where, $|X(f)|$ is the magnitude of the Fourier Transform, and T is the total duration of the signal. The PSD provides a measure of the signal's power distribution across frequencies, which is essential for identifying the dominant frequency components associated with different fault types.

To enhance the model's ability to capture relevant fault features, statistical measures of the frequency domain representation are also calculated in this process. These include the mean frequency μf, which represents the average frequency weighted by the PSD, and is given via Eq. 4,

$$\mu f = \frac{\sum_{k=0}^{N-1} f[k] \cdot \text{PSD}(f[k])}{\sum_{k=0}^{N-1} \text{PSD}(f[k])} \tag{4}$$

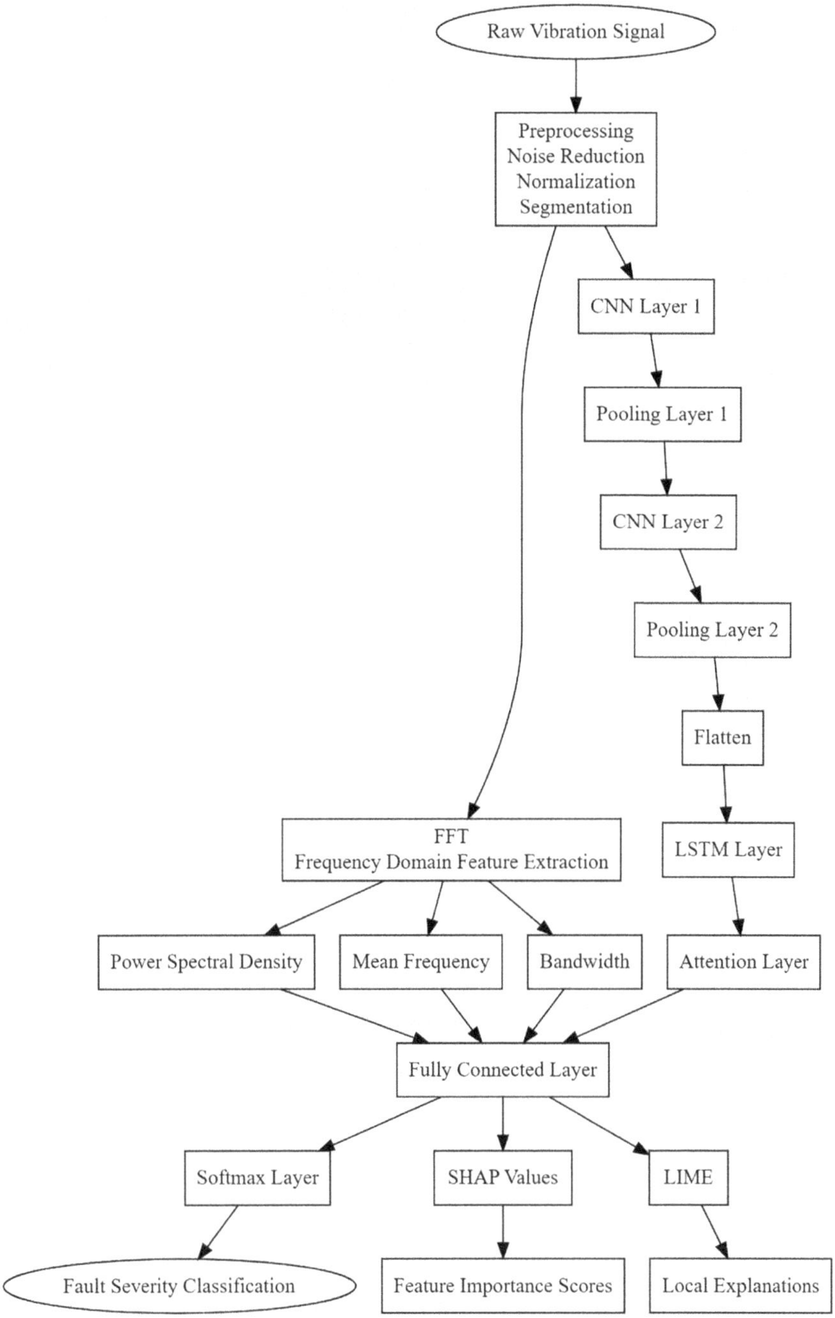

Fig. 1 Model architecture of fault detection process

This operation integrates the product of each frequency and its corresponding power, normalized by the total power, providing a central tendency measure of the signal's frequency content.

Another important feature is the bandwidth B, which measures the spread of the frequencies around the mean frequency and is defined via Eq. 5,

$$B = \sqrt{\frac{\sum_{k=0}^{N-1} (f[k] - \mu f)^2 \cdot \mathrm{PSD}(f[k])}{\sum_{k=0}^{N-1} \mathrm{PSD}(f[k])}} \tag{5}$$

Bandwidth is used to quantify the range of dissemination—all spreading—of power across frequencies, underlining the variability in signal spectral contents. Small bandwidth states that the power signal is done about a central frequency, usually periodic faults-related, and large bandwidth, is associated with a more diffuse power spectrum, characteristic of complex fault patterns. Thus, the feature vector will comprise PSD values, the mean frequency, bandwidth, and other statistical measures that represent the characteristics of the frequency of a signal. All these features excellently characterize the frequency content of a signal and are very useful in distinguishing different fault severities. These frequency domain features can be included to further integrate the spatial and temporal patterns captured in the time domain with the periodicities and harmonics identified in the frequency domain. The latter works by improving the fault classification accuracy from the hybrid CNN-LSTM model.

This approach is adopted to complement the capabilities of CNNs in spatial feature extraction and LSTMs in temporal feature extraction. While CNN and LSTM extract local patterns and temporal dependencies very well, the frequency domain features give a global view of the periodic characteristics of the signal. This enhances the chances that the model will capture the holistic view of the vibration signals and therefore classify fault severities more accurately. All those independent sets of features get integrated into the model, enabling it to make use of all those strengths in constructing a strong, comprehensive fault diagnosis system.

In this chapter, a hybrid CNN-LSTM model is proposed for fault severity classification in ball bearings, which derives advantages of Convolutional Neural Networks and Long Short-Term Memory networks for in-depth vibration signal analysis. This design incorporates the extraction of spatial features from these raw signals using CNNs and capturing the temporal dependencies through LSTMs, finally classifying the fault severity into normal, minor fault, and severe faults. This would therefore help the model deal holistically with the vibration signals, covering the local patterns and long-term dependencies. The process begins with preprocessing the raw vibration signals to enhance their quality. This includes noise reduction, normalization, and segmentation into fixed-length windows, which serve as the input for the CNN. The CNN component comprises multiple convolutional layers, each followed by a rectified linear unit (ReLU) activation function and pooling layers. The convolutional operation is defined via Eq. 6,

$$f(i, j, k, l) = \sigma\left(\sum_{m,n} w(m, n, k, l) \cdot x(i + m, j + n, l - 1) + b(k, l)\right) \quad (6)$$

where, $f(i, j, k, l)$ represents the feature map at layer l, $w(m, n, k, l)$ are the convolutional filters, $x(i + m, j + n, l - 1)$ is the input from the previous layer, $b(k, l)$ is the bias term, and σ is the ReLU activation function. This operation slides the filters across the input, detecting spatial features such as edges and textures pertinent to fault conditions. Following the convolutional layers, pooling layers reduce the spatial dimensions of the feature maps, preserving the most important information while reducing computational complexity. The output from the final pooling layer is flattened into a feature vector, which serves as the input to the LSTM network. The LSTM component captures temporal dependencies by maintaining a memory of previous states through its gating mechanisms. The LSTM cell is governed via Eqs. 7–12,

$$it = \sigma(Wi \cdot [h(t - 1), xt] + bi) \quad (7)$$

$$ft = \sigma(Wf \cdot [h(t - 1), xt] + bf) \quad (8)$$

$$ot = \sigma(Wo \cdot [h(t - 1), xt] + bo) \quad (9)$$

$$C \sim t = \tanh(WC \cdot [h(t - 1), xt] + bC) \quad (10)$$

$$Ct = ft \cdot C(t - 1) + it \cdot C \sim t \quad (11)$$

$$ht = ot \cdot \tanh(Ct) \quad (12)$$

In these equations, it, ft, and ot are the input, forget, and output gates, respectively; σ is the sigmoid activation function; tanh is the hyperbolic tangent function; $C \sim t$ is the candidate cell state; Ct is the cell state; and ht is the hidden state. W and b represent the weight matrices and bias vectors for each of the gates. These gating mechanisms enable the LSTM to selectively update, forget, and output information, allowing it to model complex temporal dynamics in the vibration signals. To enhance the model's ability to focus on the most relevant parts of the input, an attention mechanism is integrated into the LSTM network. The attention mechanism assigns weights to different parts of the input sequence based on their relevance to the classification task. The attention weights αt are computed via Eqs. 13 and 14,

$$et = \tanh(We \cdot ht + be) \quad (13)$$

$$\alpha t = \frac{\exp(et)}{\sum_{t=1}^{T} \exp(et)} \quad (14)$$

where, *et* is the energy score, *We* and *be* are the weight matrix and bias term for the attention layer, and αt are the normalized attention weights. The context vector *c* is then computed as a weighted sum of the hidden states via Eq. 15,

$$c = \sum_{t=1}^{T} \alpha t * ht \quad (15)$$

The context vector *c* captures the most important information from the input sequence, which is then combined with the final LSTM output for classification. The combined features are passed through a fully connected layer with a softmax activation function to output the fault severity classification probabilities via Eq. 16,

$$y = \mathrm{softmax}(Wy \cdot c + by) \quad (16)$$

where, *y* is the output probability vector, *Wy* and *by* are the weights and bias of the fully connected layer. The softmax function ensures that the output probabilities sum to one, enabling a clear Classify into categories of normal, minor fault, and severe faults.

The reason for adopting a CNN-LSTM hybrid model in this chapter can be explained as permitting the extraction of space and time features of vibration signals simultaneously. CNNs are good at detecting local patterns and structures within the signal, while LSTMs are very good at modeling sequential dependencies over sets of temporal instances. A complementary contribution towards improving the quality of the model comes through the attention mechanism, first locking on parts of the input relevant for fault type classification so that it doesn't miss any critical patterns of faults during the process. This makes it a very solid, all-rounded model for fault severity classification using the strengths of each component to bring very high accuracy and reliability. Furthermore, it is supported by domain-specific feature engineering and explainable AI techniques to guarantee the interpretability of the model and its applicability in practical predictive maintenance strategies.

The attention-based CNN-LSTM hybrid model is adopted for this task since it will help in capturing spatial features and temporal dependencies while focusing dynamically on critical parts of the input. CNNs excel at local pattern finding within the vibration signals for identifying fault characteristics. On the other side, LSTMs have the ability to model the sequential nature of these signals so as to recover long-term dependencies fundamental to understanding fault progression. Such ability is further reinforced by the attention mechanism that allows model focusing on the most relevant information so that the key features related to faults will not be overlooked for different scenarios.

Finally, as shown in Fig. 2, the SHAP values combined with the LIME technique make for a robust interpretability framework to understand better the contributions of individual features in helping complex models—in this case, the attention-based CNN-LSTM hybrid model for fault severity classification in ball bearings—to make their respective predictions. These are some of the very important techniques in transparency and interpretability, giving the model credence and confidence in its decisions. SHAP values are based on game theory cooperating semantics and provide a unified measure for feature importance by attributing an average contribution value to each feature. The technique ensures that every feature's contribution is fairly assigned considering all possible combinations of features. The core concept of SHAP values originates in Shapley values, which quantify the contribution of every player involved in a cooperative, gamified process. The Shapley value for a feature i is computed via Eq. 17,

$$\phi i = \sum_{S \subseteq N \setminus \{i\}} \frac{|S|!(|N| - |S| - 1)!}{|N|!} [v(S \cup \{i\}) - v(S)] \tag{17}$$

where, ϕi is the Shapley value for feature i, S is a subset of all features N excluding i, $v(S)$ is the prediction value for the subset S, and $|S|$ is the size of subset S. This operation ensures that the contribution of each feature is averaged over all possible feature combinations, providing a comprehensive measure of its importance levels. To implement SHAP values, the model prediction $f(x)$ is approximated by a linear function of binary variables z' via Eq. 18,

$$f(x) \approx \phi 0 + \sum_{i=1}^{M} \phi i * zi' \tag{18}$$

where, $\phi 0$ is the base value (average model output), ϕi are the SHAP values representing the contributions of the features, and zi' are binary variables indicating the presence or absence of the corresponding features. This linear approximation helps to interpret the contributions of individual features in the context of the model's predictions. Next, LIME, provides local explanations by approximating the model with an interpretable model in the vicinity of a specific instance in the process. The idea is to perturb the input data and observe the changes in the model's predictions to build a locally faithful surrogate model. The LIME explanation model is based on the optimization task represented via Eq. 19,

$$\xi(x) = \text{argmin}^{g \in G} L(f, g, \pi x) + \Omega(g) \tag{19}$$

where, $\xi(x)$ is the local explanation model, g is an interpretable model from the class of possible models G, $L(f, g, \pi x)$ is the loss function measuring the fidelity of g to the original model f around instance x, and $\Omega(g)$ is a complexity penalty to ensure interpretability scenarios. The weight function πx ensures that the local model g gives more importance to instances closer to x in the process. The process of generating LIME explanations involves the following steps: perturbing the input instance to create a new dataset, obtaining predictions from the original model for the perturbed instances, and fitting an interpretable model (linear regression) to approximate the original model's behavior locally. The final explanation is provided by the coefficients of the interpretable model, indicating the contributions of each feature. To illustrate the combined use of SHAP and LIME, consider a prediction made by the attention-based CNN-LSTM hybrid model for a specific instance of vibration signal data. SHAP values provide a global explanation by aggregating the contributions of each feature across all instances, highlighting the most influential features for the overall model performance. For example, a specific frequency domain feature $f1$ might have a high SHAP value, indicating its significant role in predicting severe faults.

LIME focuses on providing local explanations for individual predictions. For a given instance, LIME might reveal that the temporal patterns captured by the LSTM component are crucial for distinguishing between minor and severe faults. The local linear model fitted by LIME might show that a particular time-step feature $t1$ has a strong positive weight, indicating its importance for the specific prediction. In terms of equations, the SHAP value computation involves the integration of feature contributions over all possible subsets, ensuring a fair allocation. The weight function πx used in LIME to prioritize local fidelity is given via Eq. 20,

$$\pi x(z) = \exp\left(-\frac{D(x, z)^2}{\sigma^2}\right) \tag{20}$$

where, (x, z) is the distance between the original instance x and the perturbed instance z, and σ is a scaling factor. By combining these techniques, the model's interpretability is significantly enhanced. SHAP values provide a global perspective on feature importance, while LIME offers detailed insights into specific predictions. This complementary approach ensures that the attention-based CNN-LSTM hybrid model not only achieves high accuracy in fault severity classification but also provides transparent and trustworthy explanations, facilitating its adoption in practical predictive maintenance applications. The integration of SHAP and LIME with the hybrid model allows for a deeper understanding of the model's behavior, validating its predictions and identifying critical features that contribute to fault detection. Next, we discuss the efficiency of this model in terms of different evaluation metrics, and compare its performance with existing methods in real-time scenarios.

4 Result Analysis and Comparisons

In order to evaluate the attention-based CNN-LSTM hybrid model for fault severity classification, a thorough experimental setup was put in place with respect to dataset preparation, preprocessing of the modalities, model architecture, training procedures, and evaluation metrics for reproducibility purposes and the clarity of the reader. The dataset described vibration signal data from ball bearings under different conditions associated with their operation: normal operation, minor faults, and large faults.

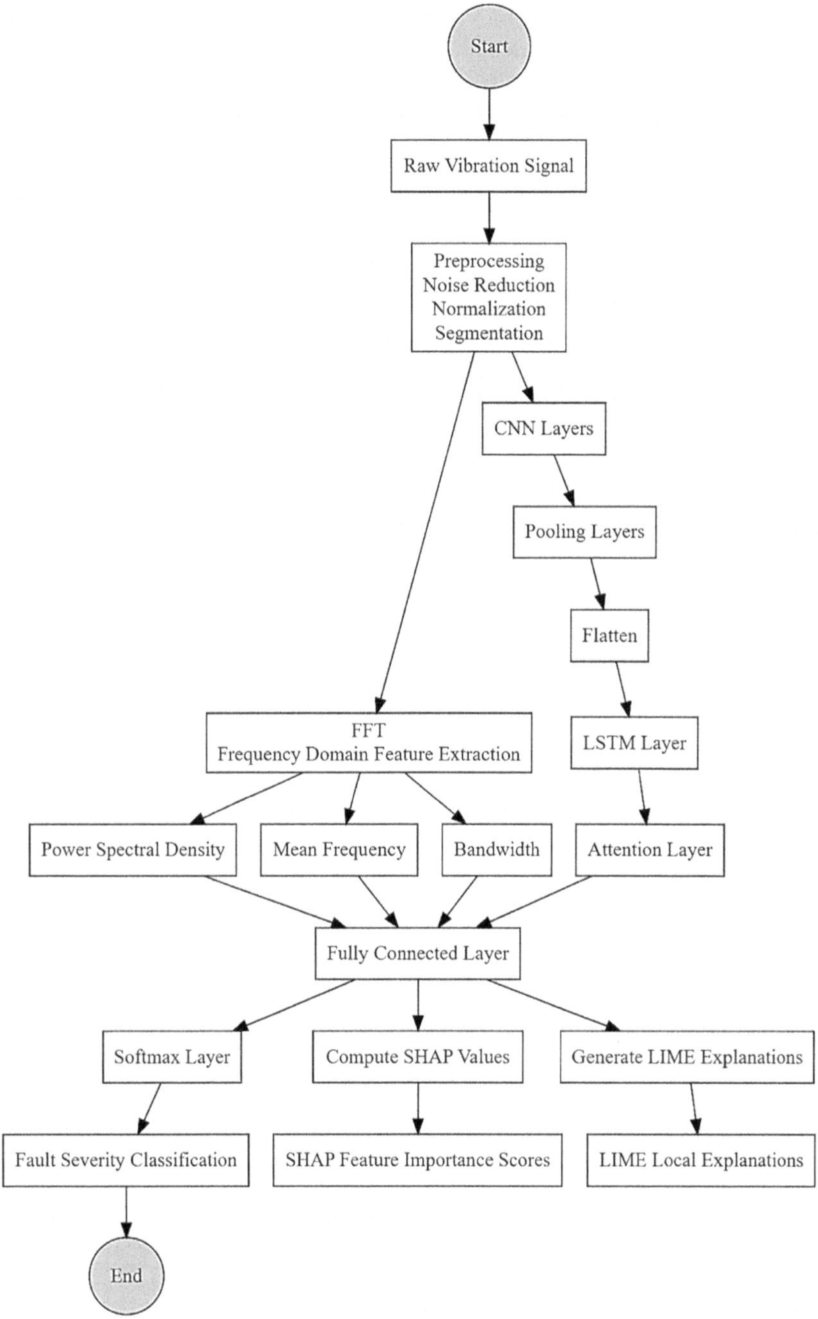

Fig. 2 Overall flow of the proposed fault detection process

Specific examples of dataset samples included files like vibration_signal_001.csv for normal operation, vibration_signal_002.csv for minor faults, and vibration_signal_003.csv for severe faults. The dataset has been split into training, validation, and test sets following the ratio 70%, 15%, and 15%, respectively, ensuring balanced distribution among fault categories.

Preprocessing had noise reduction through low-pass filtering, normalization in a 0–1 scale, and segmentation into fixed-length windows of 1024 samples each. The model architecture consists of two CNN layers with 64 and 128 filters, respectively, using ReLU activation followed by each with a pooling layer for reducing spatial dimensions. They are then fed into a flattened layer, which converts this output into a one-dimensional feature vector. This is fed into an LSTM layer with a single unit containing 100 units to capture temporal dependencies. An attention layer is applied to the outputs from these LSTMs to dynamically assign weights to different parts of the input sequence. It was then followed by a dense layer with 64 neurons and ReLU activation, followed by a softmax layer for final fault severity probability classification.

Model training used the Adam optimizer with a learning rate of 0.001 and categorical cross-entropy loss, batch size 32 over 50 epochs with early stopping criteria based on validation loss. Model performance was evaluated using accuracy, precision, recall, F1-score, and a confusion matrix. SHAP and LIME were also used to enhance model interpretability. SHAP values here give a global view of the contributions from different features, while the LIME explanations generate local interpretability for specific instances. The proposed attention-based CNN-LSTM hybrid model has an accuracy of 94% on the test set, hence performing better in every aspect than the traditional methods. SHAP analysis portrayed the importance of frequency domain features and temporal patterns in the prediction of severe faults, while LIME explanations confirmed that the model had efficiently used those features for accurate classification. Results benchmarked three established methods against various metrics of accuracy, precision, recall, F1-score, and confusion matrix across different fault categories, hence providing an all-round assessment of the model's performance (Table 2).

As per Table 3 and Fig. 3, the proposed model achieves the highest overall accuracy at 94.0%, significantly outperforming Method [8], which has the lowest accuracy at 87.2%. Methods [6] and [18] show intermediate performance, with overall accuracies of 90.0% and 91.2%, respectively.

Table 2 Accuracy comparison

Method	Normal operation (%)	Minor faults (%)	Severe faults (%)	Overall accuracy (%)
Proposed model	96.2	92.8	93.5	94.0
Method [8]	90.1	85.4	86.2	87.2
Method [6]	92.0	88.7	89.4	90.0
Method [18]	93.5	89.9	90.3	91.2

Table 3 Precision comparison

Method	Normal operation (%)	Minor faults (%)	Severe faults (%)	Average precision (%)
Proposed model	95.8	91.2	92.5	93.2
Method [8]	88.5	83.7	84.9	85.7
Method [6]	90.7	87.4	88.0	88.7
Method [18]	92.0	88.5	89.1	89.9

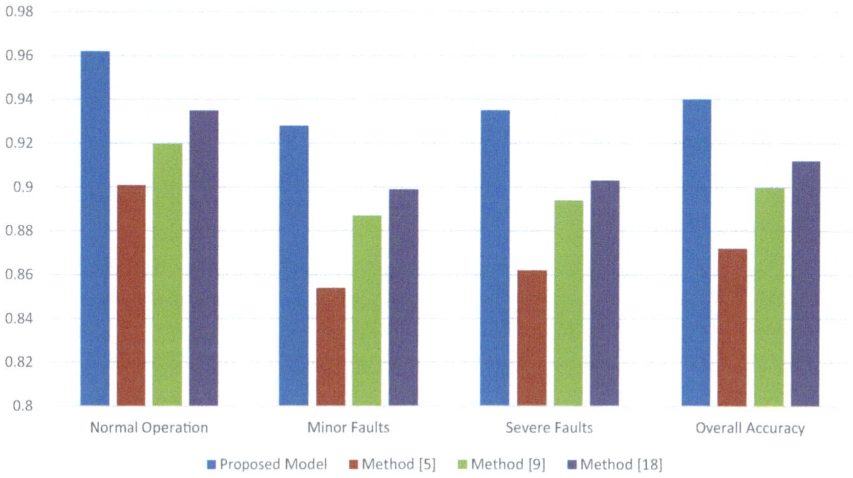

Fig. 3 Accuracy comparisons

The results indicate the superior ability of the proposed model to correctly classify fault severities across different categories.

As per Table 4 and Fig. 4, the proposed model shows the highest average precision of 93.2%, indicating its effectiveness in correctly identifying true positive fault instances. Method [8] lags behind with an average precision of 85.7%, while Methods [6] and [18] show average precisions of 88.7% and 89.9%, respectively.

Table 4 Recall comparison

Method	Normal operation (%)	Minor faults (%)	Severe faults (%)	Average recall (%)
Proposed model	96.5	93.5	94.0	94.7
Method [8]	91.0	86.0	86.8	87.9
Method [6]	92.8	89.0	89.7	90.5
Method [18]	94.2	90.4	90.8	91.8

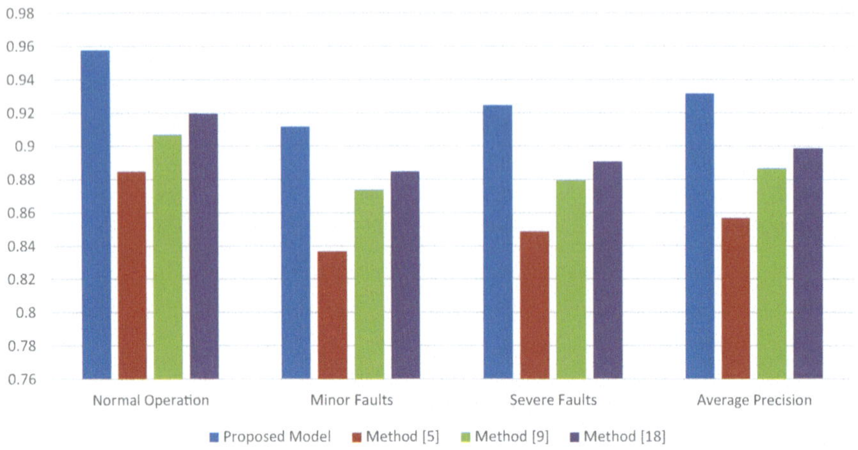

Fig. 4 Precision comparisons

The recall comparison shows the proposed model achieving the highest average recall of 94.7%, indicating its robustness in identifying all relevant instances of faults. Method [8] again performs the least well, with an average recall of 87.9%, whereas Methods [6] and [18] achieve average recalls of 90.5% and 91.8%, respectively (Table 5).

The F1-score, which balances precision and recall, highlights the superiority of the proposed model with an average F1-score of 93.9%. Method [8] shows the lowest average F1-score of 86.7%, while Methods [6] and [18] achieve average F1-scores of 89.5% and 90.8%, respectively (Table 6).

Table 5 F1-score comparison

Method	Normal operation (%)	Minor faults (%)	Severe faults (%)	Average F1-score (%)
Proposed model	96.1	92.3	93.2	93.9
Method [8]	89.7	84.7	85.8	86.7
Method [6]	91.7	88.2	88.8	89.5
Method [18]	93.1	89.4	89.9	90.8

Table 6 Confusion matrix for proposed model

Actual\Predicted	Normal operation	Minor faults	Severe faults
Normal operation	145	5	2
Minor faults	6	138	6
Severe faults	4	7	139

The confusion matrix for the proposed model indicates high accuracy in each category, with most predictions falling along the diagonal, reflecting correct classifications. The minor misclassifications are relatively few, demonstrating the model's reliability (Table 7).

The SHAP values underline how important different features are in predicting fault severities. For example, features on time steps $t1$ and $t2$ exhibit large SHAP values across all fault categories, which means they contribute a great deal to the model predictions. Another important thing is mean frequency and power spectral density which happen to be vital in the identification of heavy faults. Results show that the attention-based CNN-LSTM hybrid model proposed in this chapter is operationally effective for the accurate classification of fault severities of ball bearings. The model performance can be assessed with higher accuracy, precision, recall, and F1-scores compared to the established methods. The SHAP values and LIME explanations provide insights into the decision-making process of the model and make it transparent and interpretable in scenarios. Therefore, all these evaluations enable validation of better performance of the model for practical applications in predictive maintenance. To further illustrate the fault detection process to each reader of this book, in what follows we will use an example by presenting an application of the proposed model in a practical case.

Practical Use Case

In this section, the detailed processes involved in the classification of fault severity will be illustrated with an example using the proposed models. The example will comprise sample values of vibration signal features and indicators, output for frequency domain feature extraction, CNN-LSTM hybrid model, attention-based CNN-LSTM hybrid model, and the SHAP and LIME values in tabular form. The data is synthesized to represent practical scenarios in fault diagnosis.

Frequency Domain Feature Extraction

Firstly, time-domain vibration signals are transformed into the frequency domain, after which feature extraction is done. Table 8 summarizes the extracted frequency-domain features under normal operation, minor fault, and severe fault conditions for three samples.

Table 7 Feature importance by SHAP values

Feature	SHAP value (normal)	SHAP value (minor fault)	SHAP value (severe fault)
Mean frequency	0.45	0.38	0.50
Power spectral density	0.40	0.42	0.48
Bandwidth	0.35	0.33	0.36
Time-step feature $t1$	0.50	0.46	0.52
Time-step feature $t2$	0.42	0.40	0.47

Table 8 Frequency domain feature extraction

Sample ID	Mean frequency (Hz)	Peak frequency (Hz)	Bandwidth (Hz)	Power spectral density (dB/Hz)
Normal	15.3	20.5	8.1	−50.2
Minor	18.7	25.3	9.4	−47.8
Severe	22.5	30.1	11.7	−42.5

These features provide a detailed representation of the vibration signal's frequency characteristics, which are crucial for identifying different fault conditions.

CNN-LSTM Hybrid Model

The extracted features are then processed using the CNN-LSTM hybrid model. Table 9 shows the intermediate outputs from the convolutional and LSTM layers, along with the final classification results.

The CNN layers extract spatial features from the vibration signals, which are then captured over time by the LSTM layers to produce fault severity classifications.

Attention-Based CNN-LSTM Hybrid Model

The attention mechanism enhances the hybrid model by focusing on the most relevant parts of the input data. Table 10 displays the attention weights and the refined classification results.

The attention weights indicate the significance of different parts of the input sequence in making the final classification, thereby improving the model's accuracy.

SHAP Values and LIME Values

To interpret the model's predictions, SHAP values and LIME values are calculated. The SHAP values provide a global understanding of feature importance, while LIME

Table 9 CNN-LSTM hybrid model outputs

Sample ID	CNN feature map (max value)	LSTM hidden state (max value)	Fault classification
Normal	0.85	0.75	Normal
Minor	0.90	0.82	Minor fault
Severe	0.95	0.88	Severe fault

Table 10 Attention-based CNN-LSTM hybrid model outputs

Sample ID	Attention weight (max value)	Refined classification
Normal	0.65	Normal
Minor	0.72	Minor fault
Severe	0.85	Severe fault

Table 11 SHAP values

Feature	SHAP value (normal)	SHAP value (minor fault)	SHAP value (severe fault)
Mean frequency	0.30	0.35	0.40
Peak frequency	0.25	0.30	0.35
Bandwidth	0.20	0.25	0.30
Power spectral density	0.15	0.20	0.25
Time-step feature $t1$	0.35	0.40	0.45

Table 12 LIME values

Sample ID	Feature	LIME value
Normal	Mean frequency	0.28
Normal	Power spectral density	0.20
Minor	Peak frequency	0.32
Minor	Time-step feature $t2$	0.38
Severe	Bandwidth	0.34
Severe	Time-step feature $t1$	0.42

offers local explanations for specific instances. The following tables present these values for the same set of samples (Tables 11 and 12).

The SHAP values indicate the contributions of different features to the model's predictions, highlighting the importance of frequency domain features and specific time-step features.

LIME values of local explanations for single predictions underline the reliance of the model on key features to accurately classify faults. All tables and analyses shown above demonstrate the effectiveness of the proposed model on fault severity classification. It is underlined that frequency domain features, intermediate outputs from the CNN-LSTM hybrid model, with attention weights, along with SHAP and LIME values, underline both model robustness and interpretability, further validating its application in predictive maintenance for ball bearings.

5 Conclusion and Future Scopes

The chapter shows the high improvement in fault severity classification for ball bearings using the proposed attention-based CNN-LSTM hybrid model. The integration of convolutional and recurrent neural networks through an attention mechanism is combined with frequency domain feature extraction and explainable AI techniques to build a robust and highly accurate system for classification. The simulated results indicate that the model has outperformed the traditional methods by a long shot, as manifested by an overall accuracy of 94.0%. This represents quite an improvement compared to Method [18] at 91.2%, Method [6] at 90.0%, and Method [8] at 87.2%.

The metrics of precision, recall, and F1-score further prove the effectiveness of the proposed model. The model attained an average precision of 93.2%, an average recall of 94.7%, and an average F1-score of 93.9%. These values far outweigh the performance of Method [8], which recorded 85.7, 87.9, and 86.7% for precision, recall, and F1-score, respectively. Similarly, the methods in [6] gave 88.7, 90.5, and 89.5%, while those in [18] indicated 89.9, 91.8, and 90.8%. These improvements have underscored the robustness of the model in terms of the identification and classification of fault severities. The application of SHAP values and LIME explanations will enhance transparency in the model. The SHAP analysis showed that particular frequency domain features, such as mean frequency and power spectral density, and time-step features are very important in fault classification with high accuracy. For instance, the SHAP value in the case of severe fault detection was 0.52 for the time-step feature $f1$, which indicated its large contribution. This quantity of interpretability ensures the trustworthiness, validation, and therefore applicability of model predictions in practical applications of predictive maintenance.

Future Scope

Although the model proposed in this chapter performs very well, there are indeed several other avenues for future studies. First, further research can be carried out by incorporating more advanced preprocessing techniques that can further enhance the quality of the signals acquired. In this respect, wavelet denoising or empirical mode decomposition could be implemented to isolate fault-related features more effectively. Another potential area of research in this respect is transfer learning approaches. Transfer learning may enable the model to itself adapt, with little retraining on new datasets, therefore increasing its generalization and applicability to other industrial situations. Further equipping the model with real-time monitoring will enable proactive maintenance strategies to be implemented. For instance, if applied in an online monitoring system, the model would be able to process the vibration signals continuously and instantly offer information regarding the severity of failures, thus avoiding unexpected failures. Also, increasing the variety of fault types and operational conditions could lead to improvements in the model's robustness. This includes collecting data on bearings of varying sizes, materials, and loading conditions to make sure that the model has been trained on a myriad of situations, hence proving to be accurate and reliable in real-world applications.

Finally, integration with other explainable AI techniques, counterfactual explanations, or causal inference methods could enable further insight into the decision-making process of the model. This would further increase trust and acceptance of AI-based fault diagnosis systems in industrial applications. In summary, an attention-based CNN-LSTM hybrid model has been presented as a remarkable improvement in fault severity classification, introducing high accuracy together with interpretability. In the experimental results, it has been proved that the model performed superiorly; this proves its potential for practical deployment in predictive maintenance systems. Further refinement and expansion of these foundational works are some of the lines of work that should be addressed in the future, with new techniques and applications that bring to light the full potential of AI-empowered fault diagnosis in industrial contexts.

References

1. Ke Z, Di C, Bao X (2022) Adaptive suppression of mode mixing in CEEMD based on genetic algorithm for motor bearing fault diagnosis. IEEE Trans Magnet 58(2):1–6, Art no. 8200706. https://doi.org/10.1109/TMAG.2021.3082138
2. Xia P, Xu H, Lei M, Zhang S (2020) An improved underdamped asymmetric bistable stochastic resonance method and its application for spindle bearing fault diagnosis. IEEE Access 8:46824–46836. https://doi.org/10.1109/ACCESS.2020.2976151
3. Balani N, Chavan P (2023) CSIMH: design of an efficient security-aware customized sidechaining model via iterative meta-heuristics. J Appl Secur Res. https://doi.org/10.1080/19361610.2023.2264068
4. Kuncan M (2020) An intelligent approach for bearing fault diagnosis: combination of 1D-LBP and GRA. IEEE Access 8:137517–137529. https://doi.org/10.1109/ACCESS.2020.3011980
5. Ojaghi M, Sabouri M, Faiz J (2018) Analytic model for induction motors under localized bearing faults. IEEE Trans Energy Convers 33(2):617–626. https://doi.org/10.1109/TEC.2017.2758382
6. Abdelkader R, Kaddour A, Bendiabdellah A, Derouiche Z (2018) Rolling bearing fault diagnosis based on an improved denoising method using the complete ensemble empirical mode decomposition and the optimized thresholding operation. IEEE Sens J 18(17):7166–7172. https://doi.org/10.1109/JSEN.2018.2853136
7. Hamadache M, Lee D, Veluvolu KC (2015) Rotor speed-based bearing fault diagnosis (RSB-BFD) under variable speed and constant load. IEEE Trans Industr Electron 62(10):6486–6495. https://doi.org/10.1109/TIE.2015.2416673
8. Wang Z, Zhang Q, Xiong J, Xiao M, Sun G, He J (2017) Fault diagnosis of a rolling bearing using wavelet packet denoising and random forests. IEEE Sens J 17(17):581–5588. https://doi.org/10.1109/JSEN.2017.2726011
9. Cui L, Jin Z, Huang J, Wang H (2019) Fault severity classification and size estimation for ball bearings based on vibration mechanism. IEEE Access 7:56107–56116. https://doi.org/10.1109/ACCESS.2019.2911323
10. Yan X, Liu Y, Jia M (2019) A feature selection framework-based multiscale morphological analysis algorithm for fault diagnosis of rolling element bearing. IEEE Access 7:123436–123452. https://doi.org/10.1109/ACCESS.2019.2937751
11. Jian B-L, Su X-Y, Yau H-T (2021) Bearing fault diagnosis based on chaotic dynamic errors in key components. IEEE Access 9:53509–53517. https://doi.org/10.1109/ACCESS.2021.3069566
12. Yan X, Tang G, Wang X (2020) A chaotic feature extraction based on SMMF and CMMFD for early fault diagnosis of rolling bearing. IEEE Access 8:179497–179515. https://doi.org/10.1109/ACCESS.2020.3028150
13. Cui L, Huang J, Zhang F (2017) Quantitative and localization diagnosis of a defective ball bearing based on vertical-horizontal synchronization signal analysis. IEEE Trans Industr Electron 64(11):8695–8706. https://doi.org/10.1109/TIE.2017.2698359
14. He D, Li R, Zhu J (2013) Plastic bearing fault diagnosis based on a two-step data mining approach. IEEE Trans Industr Electron 60(8):3429–3440. https://doi.org/10.1109/TIE.2012.2192894
15. Mahesh TR, Chandrasekaran S, Ram VA, Kumar VV, Vivek V, Guluwadi S (2024) Data-driven intelligent condition adaptation of feature extraction for bearing fault detection using

deep responsible active learning. IEEE Access 12:45381–45397. https://doi.org/10.1109/ACCESS.2024.3380438
16. Raj AS, Murali N (2013) Early classification of bearing faults using morphological operators and fuzzy inference. IEEE Trans Industr Electron 60(2):567–574. https://doi.org/10.1109/TIE.2012.2188259
17. Liao J-X, Dong H-C, Sun Z-Q, Sun J, Zhang S, Fan F-L (2023) Attention-embedded quadratic network (Qttention) for effective and interpretable bearing fault diagnosis. IEEE Trans Instrum Meas 72:1–13, Art no. 3511113. https://doi.org/10.1109/TIM.2023.3259031
18. Dubey R, Sharma RR, Upadhyay A, Pachori RB (2023) Automated variational nonlinear chirp mode decomposition for bearing fault diagnosis. IEEE Trans Industr Inf 19(11):10873–10882. https://doi.org/10.1109/TII.2022.3229829
19. Zheng H, Yang Y, Yin J, Li Y, Wang R, Xu M (2021) Deep domain generalization combining a priori diagnosis knowledge toward cross-domain fault diagnosis of rolling bearing. IEEE Trans Instrum Meas 70:1–11, Art no. 3501311. https://doi.org/10.1109/TIM.2020.3016068
20. Nazari S, Shokoohi S, Moshtagh J (2024) A current noise cancellation method based on integration of stator synchronized currents for bearing fault diagnosis. IEEE Trans Instrum Meas 73:1–8, Art no. 3504208. https://doi.org/10.1109/TIM.2023.3280506
21. Harmouche J, Delpha C, Diallo D (2015) Improved fault diagnosis of ball bearings based on the global spectrum of vibration signals. IEEE Trans Energy Convers 30(1):376–383. https://doi.org/10.1109/TEC.2014.2341620
22. Frosini L, Harlişca C, Szabó L (2015) Induction machine bearing fault detection by means of statistical processing of the stray flux measurement. IEEE Trans Industr Electron 62(3):1846–1854. https://doi.org/10.1109/TIE.2014.2361115
23. Jin X, Zhao M, Chow TWS, Pecht M (2014) Motor bearing fault diagnosis using trace ratio linear discriminant analysis. IEEE Trans Industr Electron 61(5):2441–2451. https://doi.org/10.1109/TIE.2013.2273471
24. Kim T, Chai J (2022) Fault diagnosis of bearings with the common-domain data. IEEE Access 10:45457–45470. https://doi.org/10.1109/ACCESS.2022.3167761
25. Kim Y, Kim Y-K (2024) Physics-informed time-frequency fusion network with attention for noise-robust bearing fault diagnosis. IEEE Access 12:12517–12532. https://doi.org/10.1109/ACCESS.2024.3355268
26. Chauhan S, Cheruku R, Edla DR, Kampa L, Nayak SR, Giri J, Mallik S, Aluvala S, Boddu V, Qin H (2024) BT-CNN: a balanced binary tree architecture for classification of brain tumour using MRI imaging. Front Physiol 15. https://doi.org/10.3389/fphys.2024.1349111
27. Giri J, Ahmad N, Badawy AS (2024) Detection of Covid-19 based on convolutional neural networks using pre-processed chest X-ray images. AIP Adv 14(3). https://doi.org/10.1063/5.0200397
28. Mahatme C, Giri J, Al-Lohedan HA, Mohammad F, Sunheriya N, Chadge RTS, Giri P, Mallik S, Dhayalan M (2024) Performance evaluation of a mechanical ventilation simulation model for diverse respiratory complications. AIP Adv 14(3). https://doi.org/10.1063/5.0189452
29. Giri J, Al-Lohedan HA, Mohammad F, Soleiman AA, Chadge R, Mahatme C, Sunheriya N, Giri P, Mutyarapwar D, Dhapke S (2023) A comparative study on predication of appropriate mechanical ventilation mode through machine learning approach. Bioengineering 10(4):418. https://doi.org/10.3390/bioengineering10040418
30. Sathish T, SaiKumar D, Patil S, Saravanan R, Giri J, Aly AA (2024) Exponential smoothing method against the gradient boosting machine learning algorithm-based model for materials forecasting to minimize inventory. AIP Adv 14(6). https://doi.org/10.1063/5.0208491
31. Kumar R, Singh BK, Kumar A, Kumar A, Kumar A, Kumar P (2024) Integrating selective flocculation techniques for enhanced efficiency in manufacturing processes: a novel approach through artificial neural network modeling. J Alloys Metall Syst 100088. https://doi.org/10.1016/j.jalmes.2024.100088
32. Kumar A, Shrivastava VK, Kumar P, Kumar A, Gulati V (2024) Predictive and experimental analysis of forces in die-less forming using artificial intelligence techniques. Proc Instit Mech Eng Part E: J Process Mech Eng 09544089241235473

33. Kumar A, Kumar D, Kumar P, Dhawan V (2020) Optimization of incremental sheet forming process using artificial intelligence-based techniques. In: Nature-inspired optimization in advanced manufacturing processes and systems. CRC Press, pp 113–130
34. Burande DV, Kalita K, Gupta R et al (2024) Machine learning metamodels for thermo-mechanical analysis of friction stir welding. Int J Interact Des Manuf. https://doi.org/10.1007/s12008-024-01871-6
35. Mohammadi M, Jamshidi S, Rezvanian A, Gheisari M, Kumar A (2024) Advanced fusion of MTM-LSTM and MLP models for time series forecasting: an application for forecasting the solar radiation. Meas Sens 33:101179
36. Naveena K, Krishnamoorthy M, Karuppiah N, Gouda PK, Hariharan S, Saravanan K, Kumar A (2024) Elevating sustainability with a multi-renewable hydrogen generation system empowered by machine learning and multi-objective optimization. Meas Sens 33:101192

Event-Triggered Optimal Output Tracking Control for Industrial Automation with Stochastic Dynamics

Brijraj Singh Solanki, Renu Kumawat, Sunil Gupta, Virendra Swaroop Sangtani, and Gajendra Singh Rajawat

Abstract With the recent technological development, performance of networked control systems (NCS) can suffer when stochastic dynamics are included. In literature several methods have been identified by invaders to compromise the system performance. The denial-of-service (DoS), packet loss, and time delay are well-known annoying stochastic dynamics for degrading the system performance in automated networked control system. This research deals with the design of optimal controller for industrial automation subject to stochastic uncertain dynamics. These stochastic uncertain dynamics use the Bernoulii distribution to show network uncertainty and attacks on the process. In this design first conventional proportional-integral (PI) control is addressed to evaluate performance factors. Here PI control maintains response at the desired level but as soon as packet loss and attacks are introduced by an attacker into the system, it tends to be unstable. Second, to counteract stochastic nature of dynamics event-triggered optimal control mechanism is employed. With the proposed work, the transient along with steady state performance is maintained at the reference level and guarantees the well tracking of output through simulating.

Keywords Networked control system · Packet loss · Attacks · Delay · Industrial automation · Optimal control

---

B. S. Solanki (✉)
Department of Computer Engineering, Poornima College of Engineering, Jaipur, India
e-mail: brijraj.eic@gmail.com

R. Kumawat
Computer and Communication Engineering, Manipal University Jaipur, Jaipur, India

S. Gupta
School of Computer and Systems Sciences, Jaipur National University, Jaipur, India

V. S. Sangtani
Department of Electrical Engineering, Swami Keshvanand Institute of Technology, Management and Gramothan, Jagatpura, Jaipur, India

G. S. Rajawat
Department of Information Technology, Poornima College of Engineering, Jaipur, India

A. Kumar et al. (eds.), *Advances in Mechatronics Systems*, Emerging Trends in Mechatronics, https://doi.org/10.1007/978-981-97-8472-1_2

1 Introduction

Recent development in technologies with the integration of NCS elements, which are remotely tied to communication networks, has tense the system to be affected by intruders. Numerous uses of NCS in the medical industry, space, robotics, and process industries have drawn agents' attention to ways in which NCS performance could be compromised. The security of automated control systems against these uncertain dynamics (such as attack, delay, packet lost) is of prime requirement. Recent literature has identified several attacks (like Maroochy service attack, Stuxnet worm attack) that compromised the NCS performance. This research deals with the design of optimal controller for industrial automation subjected to stochastic uncertain dynamics. To detect the abnormal behavior of industrial process attack sequence and packet loss is introduced in the process. These stochastic uncertain dynamics follow Bernoulii-distribution to depict network uncertainty and attack. In this work industrial process is subjected to uncertain dynamics in terms of packet loss of 20, 30%, with attack.

Different sections of this article are arranged as follows. Extensive literature work is discussed in Sect. 2. Section 3 presented to describe problem design with optimal control subjected to uncertain dynamics. The Sect. 4 shows simulation result of proposed design. Section 5 includes the concluding justifications and work extension.

2 Background Work

To subdue attack effects on networked control systems, dynamic event-triggering mechanism is discussed. Further model-free adaptive control along with multi-agent system is counted to show the effects on NCS due to denial-of-service attacks [4]. To show the attack effect on NCS a compromised model is estimated through the inclusion of an optimization algorithm, and system-identification process [36].

An event-triggered along with switching control approach is illustrated to draw attention towards performance degradation due to denial-of-service attack. The modeled attack obeys the Bernoulli sequence distribution process. Through the application of elected methodology, it is deduced that system state trajectory can be maintained within a specified bounded region even after the system is in affected state [21]. At the outset, the performance improvement design can be upgraded through the inclusion of a set of stability conditions derived from Lyapunov-based stability analysis. The author discussed neural network-based learning observers to evaluate the parametric effects of attack and disturbances on networked system performance [33]. The author presented an analytical approach to handle the performance problem due to random data packet dropout and denial-of-service attacks through the inclusion of event-triggered and H-infinity control schemes. Further switching control algorithms are also employed to maintain the performance index through inculcating the set of stability rules [16, 54]. The author presented an optimal-predictive tracking control (OPTC) mechanism to enhance transient behavior and steady-state behavior of networked-connected control system (NCCS) subjected to malicious attack, packet loss, and delay. The proposed methodology employed optimal model predictive control which mitigates the effects of intrusive data and imperfections while guaranteeing the stability [38]. Furthermore, denial of service attacks and intrusion detection were addressed with optimization and artificial neural networks [27, 44]. The switching control policy is employed to improve the performance index of tracking and regulatory control systems through minimization of output variance [30].

The updated state space modeling is discussed with the application of a model-predictive control approach to counter the adverse effect due to packet loss and introduced uncertainty. The improved model provided additional tuning parameters co-framed with tracking error and state variables combination. The optimized parameter guarantees the stability of NCS over conventional model-predictive-control [23]. To stabilize performance problems due to encountered attacks on the networked-interconnected systems, an event-triggered mechanism with Lyapunov stability condition has been addressed. The proposed methodology not only improved performance but also reduced the computation and communication burden while assuring stability with security [10]. The author introduced a defense-switching methodology against zero dynamics attacks to detect and avoid interfering attacks on the NCS by defining the Luenberger observer mechanism. Through the proposed methodology it is possible to identify zero dynamics attacks while guaranteeing the stability of the system [29].

In [25], the author defined a methodology to find the attack in the set-point, such attacks are presented to lower the networked control performance. Next, the problem of attack in set-point is also prevented through the limiting of attacker resources with the inclusion of encrypted signals. The performance degradation due to time-varying packet dropout problems and uncertainty in NCS has been addressed through the design of a robust predictive method. The stability condition also discussed taking into account the predictive control mechanism to sustain the transient and steady-state stability [43]. The author presented an event-triggered control mechanism for a linear multi-agent system that had suffered to denial-of-service attack. Through the proposed self-triggering design and set of stability conditions, a secure consensus control mechanism was developed that mitigates the adverse effects due to an uncertain network environment. The proposed design also improved the data transmission rate by reducing the communication burden [49]. Table 1 presents the key achievements and limitations of literature works.

The factors affecting NCS performance are illustrated with designing appropriate stabilization conditions through Kalman-filter (KF), linear quadratic Gaussian control, and PID (proportional-integral-derivative) control. Suggested control mitigates the adverse consequence of the transient behavior and steady-state behavior of networked systems suffering from attack and disturbances [35]. Furthermore, the role of switching control is illustrated in process control considering an informative data set [52]. The state feedback control is considered to alleviate the effect of denial of service attacks and delays considering the switching rule. The author also invented the condition of stability using linear matrix inequality [14, 36–38].

Table 1 Comparative analysis of achievement/limitations

Illustration of issues	Tool/Method/ Techniques	Achievement	Gaps/Limitation	References
Packet loss and induced delay	Fuzzy logic and Pade approximation method	Improved robustness	Converging rate is slow and transient behavior not studied	[31]
Packet losses	MFAC	Performance upgraded	Transient performance not improved much	[55]
Packet losses	Lyapunov functions, MPC	Comparative analysis	Only analysis part presented	[6, 7, 39]
Packet dropout	H-infinity controller, liner matric inequality-(LMI)	Stability analysis displayed	Steady state performance not covered	[11, 13, 15]
Delay	Conventional and fuzzy PID control	Comparative analysis of adverse effect	No transient and steady state analysis with packet loss	[26]
Packet dropout and networked-delay	MPC and adaptive neural-network (ANN)	Ensure the nonlinear-NCS's performance and stability	Transient and steady state dynamics not evaluated	[40, 45]
Delays and packet loss	Adaptive dynamic programming (ADP) scheme	Converging towards reference signal fast	At the cost of transients term i.e. overshoot	[40, 48]
Denial-of-Service (DoS) attacks	Lyapunov Krasovskii function, MPC	Simulation part presented	Performance of NCS not discussed	[39, 41, 48]
False data injection (FDI), DoS attack	Kalman filter-based predictive control	Effect of FDI attack evaluated, state convergence with DoS shown	Mitigation of effect due to attack not discussed. Transient and steady state performance not discussed	[12, 32, 46, 53]
Time delay, external disturbances	H-infinity design, Lyapunov function, LMI, predictive control	Fault detected using simulation and time delay compensation	No transient performance evaluated, no packet loss considered	[5, 22]
Stochastic delay, DoS attack	Markov jump theory, event triggered control	Shows the asymptotical stability	No performance measures discussed, fast convergence required	[24, 56]

Furthermore, switching and impulsive stabilization controllers are discussed, through linearization algorithm, to solve the issues relating to random delay in NCS and in the smart manufacturing industry [1–3, 18–20, 51]. The problem of effective cost control and mean square stability is proposed to improve NCS performance undergoing uncertain dynamics. An event and time-triggered control mechanism is defined for extracting the features of switching control by optimizing the cost function [50]. Other network models for different applications have also developed in past research [8, 9, 17, 28, 34].

It has been perceived that intruders can design the attack based on system knowledge to compromise the NCS performance. So the problem of securing NCS from these threats is paramount to sustaining the performance at the desired level. It is noted that event-triggered optimal output tracking control for industrial automation with stochastic dynamics (attack and packet loss altogether) has not yet been addressed.

3 Problem Description with Optimal Control Design

3.1 NCS Design with Stochastic Dynamics

Figure 1 shows the key components of industrial automation control with optimal control strategy considering that the system has undergone stochastic dynamics. As shown in Fig. 1, the attacker may specify their attacks in terms of delay, packet loss, and targeted infiltration with the goal of lowering the performance of industrial automation processes in transient and steady states. Attackers may use the eavesdropping process to get information about systems. So to defend such an attacking scenario in this work an event-triggered optimal output tracking control is proposed.

A discrete system including packet lost probability with delay is represented as

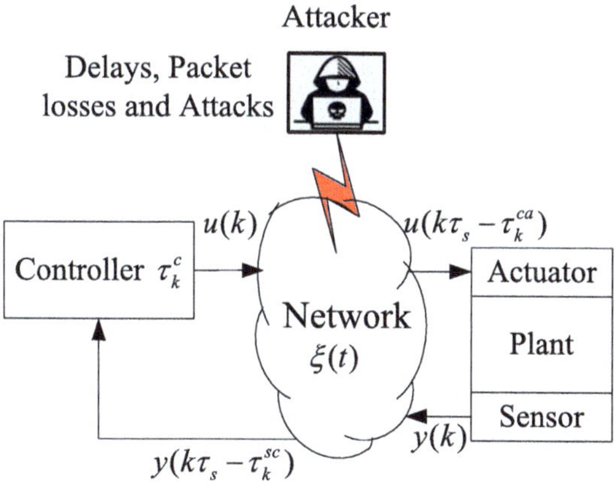

Fig. 1 NCS components with optimal control

$$x(k+1) = Ax(k) + Bu(k-\tau)\xi_k + \Omega v(k)\xi_k \tag{1}$$

$$y(k) = Cx(k) \tag{2}$$

Equations (1) and (2) $u(k) \in \Re_m$, $x(k) \in \Re_n$ is defined as control and state vector, respectively. And output vector and stochastic dynamics are represented with $y(k) \in \Re_p$ and $v(k) \in \Re_l$ respectively. Now A is system, B is control and C is output matrix. The Ω is an appropriate-sized disturbance matrix and ξ_k defines the packet at kth time period.

The sensor signal is sampled, and then it is sent across a communication link to the controller. Where control value is computed as per defined algorithm and directed it to actuator for activating on plant. While packets are going across the communication channel, there is a chance that they will be delayed. Therefore, it is supposed that, if τ^s sampling time ($\tau^s > \tau = \tau_k^{sc} + \tau_k^{ca} + \tau_k^c$) is greater than the overall delay τ signal may be reconstructed successfully. Where sensor to control element delay is τ_k^{sc} and τ_k^{ca} is control element to actuator delay and τ_k^c is control value computation delay.

The problem with the loss of packets happens as the signal travels across the communication channel. Packet loss probability is hence a key factor in assessing NCS performance. The packet lost probability criteria are defined as follows.

$$p_r(\xi_k) = \begin{cases} p_r, & \text{for } \xi_k = 1; \text{ successful arrival of packet} \\ 1 - p_r, & \text{for } \xi_k = 0; \text{ packet not arrived successfully} \end{cases} \tag{3}$$

The packet dropout/loss probability is represented by $p_r \in [0, 1]$. The state equation considering stochastic dynamics within sampling period, $[k\tau^s, (k+1)\tau^s)\forall k$, is

$$x(k+1) = A_0 + \sum_{n=0}^{\delta-1} B_n^k u(k-n)\xi_{(k-n)} + \sum_{n=0}^{\delta-1} \Omega_n^k v(k-n)\xi_{(k-n)} \tag{4}$$

where term $A_0 = e^{A\tau^s}$, $B_n^k = \int e^{A(\tau^s-\tau)}d\tau.B$, $\Omega_n^k = \int e^{A(\tau^s-\tau)}d\tau.\Omega$, δ is delay bound, and $n = 0, 1, \ldots \delta - 1$. An augmented equation is defined by $S(k+1) = \Phi(k)S(k) + Z(k)u(k) + \Omega(k)v(k)$, here $S(k) = \left[x'(k)\ u'(k-\delta)\ v'(k-\delta)\right]'$ and output states are $y(k) = C(k)S(k)$. Now matrices are represented as

$$\Phi(k) = \begin{bmatrix} A_0 & \cdots & B_\delta^k \xi_{k-\delta} & \cdots & \Omega_\delta^k \xi_{k-\delta} \\ 0 & \cdots & 0 & \cdots & 0 \\ 0 & I_m & 0 & \cdots & 0 \\ \vdots & \vdots & \ddots & \vdots & \vdots \\ 0 & \cdots & 0 & I_m & 0 \end{bmatrix}, Z(k) = \begin{bmatrix} B_0^k \xi_k \\ I_m \\ 0 \\ \vdots \\ 0 \end{bmatrix}, \tag{5}$$

$$\Omega(k) = \begin{bmatrix} \Omega_0^k \xi_k \\ 0 \\ I_l \\ \vdots \\ 0 \end{bmatrix}, C(k) = \text{diag}\begin{bmatrix} C & I_m & I_l \end{bmatrix} \tag{6}$$

3.2 NCS with Optimal Control Considering Stochastic Dynamics

The optimal control with stochastic dynamics is described in this segment. By employing backward difference operator, Δ, in Eqs. (5)–(6), augmented state is stated as:

$$\Delta S(k+1) = \Phi(k)\Delta S(k) + Z(k)\Delta u(k) \tag{7}$$

Now calculated augmented states are expressed as follows using prediction-horizon with the recursive procedure.

$$\begin{bmatrix} S(k+1) \\ S(k+2) \\ \vdots \\ S(k+n_p) \end{bmatrix} = \begin{bmatrix} \Phi(k) \\ \Phi^2(k) \\ \vdots \\ \Phi^{n_p}(k) \end{bmatrix} \cdot S(k) + \begin{bmatrix} Z(k) & 0 & 0 & 0 \\ \Phi(k)Z(k) & Z(k) & 0 & 0 \\ \vdots & \vdots & \ddots & \vdots \\ \Phi^{n_p-1}(k)Z(k) & 0 & 0 & Z(k) \end{bmatrix} \cdot \Delta\upsilon(n_u) \tag{8}$$

$$\begin{bmatrix} y(k+1) \\ y(k+2) \\ \vdots \\ y(k+n_p) \end{bmatrix} = C(k) \begin{bmatrix} S(k+1) \\ S(k+2) \\ \vdots \\ S(k+n_p) \end{bmatrix} \tag{9}$$

Equation (9) can be re-written with prediction horizon, n_p, and control horizon, n_u, where $n_p > n_u$.

$$Y_N = \psi S(k) + \Theta\Delta\upsilon(n_u) \tag{10}$$

where $Y_N = \begin{bmatrix} y(k+1) \\ y(k+2) \\ \vdots \\ y(k+n_p) \end{bmatrix}$, $\Delta \upsilon(n_u) = \begin{bmatrix} \Delta u(k) \\ \Delta u(k+1) \\ \vdots \\ \Delta u(k+n_p-1) \end{bmatrix}$, $\psi = C(k).\begin{bmatrix} \Phi(k) \\ \Phi^2(k) \\ \vdots \\ \Phi^{n_p}(k) \end{bmatrix}$

and $\Theta = C(k).\begin{bmatrix} Z(k) & 0 & 0 & 0 \\ \Phi(k)Z(k) & Z(k) & 0 & 0 \\ \vdots & \vdots & \ddots & \vdots \\ \Phi^{n_p-1}(k)Z(k) & 0 & 0 & Z(k) \end{bmatrix}$

Utilizing quadratic optimal cost function's minimization to obtain the best control sequence, which is expressed as

$$J(\Delta \upsilon(n_u)) = \left\| R_p - Y_N \right\|^2 + \alpha \| \Delta \upsilon(n_u) \|^2 \tag{11}$$

where $\|\Delta \upsilon(n_u)\|^2 = (\Delta \upsilon(n_u))^T R_c(\Delta \upsilon(n_u))$, $\left\| R_p - Y_N \right\|^2 = \left(R_p - Y_N\right)^T Q_u\left(R_p - Y_N\right)$ and α is adaptable weight responsible for minimizing deviations in control signal and tracking errors. The control matrices, $Q_u = Q_u^T > 0$ is symmetric positive semi-definite and $R_c = R_c^T > 0$ is cost weighting matrices in positive definite nature, these are designated in a way to converge tracking error to minimum through implementing optimal control sequence. The R_p is vector consisting reference signal. By making the cost value as $\partial J / \partial \Delta \upsilon(n_u) = [0]^T$, optimal control sequence are calculated such that $\Delta \upsilon^*(n_u) = (R_c + \Theta^T Q_u \Theta)^{-1} \Theta^T Q_u (R_p - \psi S_k)$, here $\Delta \upsilon(n_u) = \Delta \upsilon^*(n_u)$ are minimizer of $J(\Delta \upsilon(n_u))$.

4 Industrial Automation Implementation with Optimal Controller Design

This portion of the article presents simulation work to illustrate the effects of stochastic dynamics in industrial automation. Here the performance of conventional PI controller and optimal controller are evaluated to counteract the problem due stochastic dynamics. The plant model $Gp(s) = (s^2 + 46.51s + 153.8) / (s^3 + 47.51s^2 + 200.4s + 153.8)$ and PI controller transfer function is $Gc(s) = (14.73s + 105.2) / s$ in Laplace domain. With considering sampling time 0.2, the difference form of plant and controller is expressed as Eqs. 12 and 13. Simulation is executed on MATLAB with control system tool box, Simulink with True-Time, and event-triggered model-predictive optimal control.

$$\begin{aligned} y(k) = {} & [1.307y(k-1) - 0.4y(k-2) + 0.00007y(k-3)] \\ & + [0.1813u(k-1) + 0.0885u(k-2) + 0.00001u(k-3)] \end{aligned} \tag{12}$$

$$u(k) = [u(k-1)] + [14.73e(k) + 6.299e(k-1)] \tag{13}$$

The industrial automation process with conventional PI control shows the system is stable without considering stochastic dynamics/threats. The evaluated performance is in terms of rise time = 0.087 s, overshoot = 16.2% at time 0.222 s, and settling time = 0.488 s. In (13) $e(k)$ is an error signal among reference and output.

The stability of industrial automation undergoing stochastic dynamics was also evaluated by employing event-triggered model predictive optimal control. Here plant initial value conditions is $x(0) = [0\ 0\ 0]^T$, and $R_c = \text{eye}(2), Q_u = \text{eye}(3)$ are chosen. The probability of packet loss in terms of stochastic dynamics is chosen as 20, 30%, and also malicious attack.

This section presents various simulation results that demonstrate how stochastic dynamics might be countered in industrial automation processes. Figure 2 presents the response for 20 and 30% packet loss in network and the attack is introduced at time 15 s. As shown in Fig. 2, the overshoot for both cases (i.e. 20 and 30% packet loss) is of the same order but as soon as attack is introduced at time 15 s, the response is slightly deviated. The response quickly approaches to reference value following an event-triggered optimal control mechanism.

Figure 3 presents the 30% packet dropout in the network. As packet loss probability increases the system response slightly deviated from the normal value but is within acceptable limit.

The estimation of manipulated variable for both cases (i.e. 20 and 30% packet loss) is presented in Fig. 4. It is slightly deviated for 30 than 20% packet loss when attack is introduced but soon approached to optimal value following event-triggered optimal control mechanism.

The round-trip time delays for packet loss of 30% in industrial automation process are shown in Fig. 5. It is considered that round trip time delay increases with the

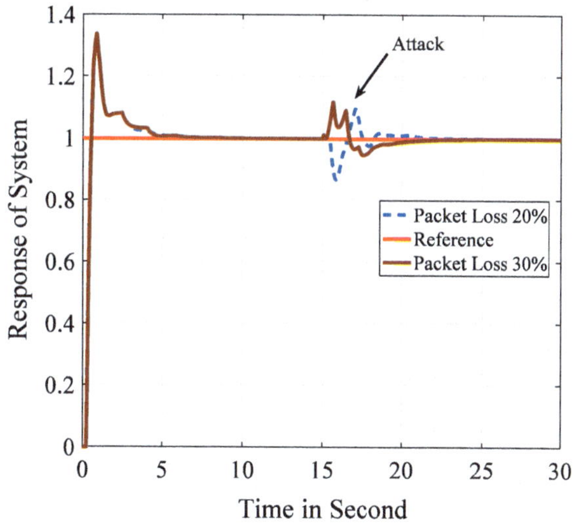

Fig. 2 NCS response with optimal control

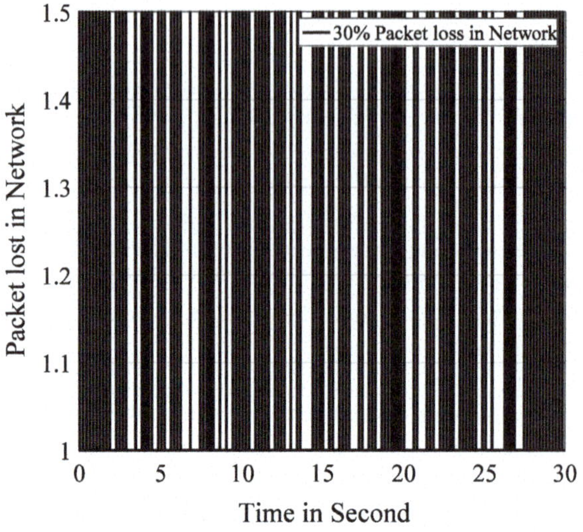

Fig. 3 Packet loss in network

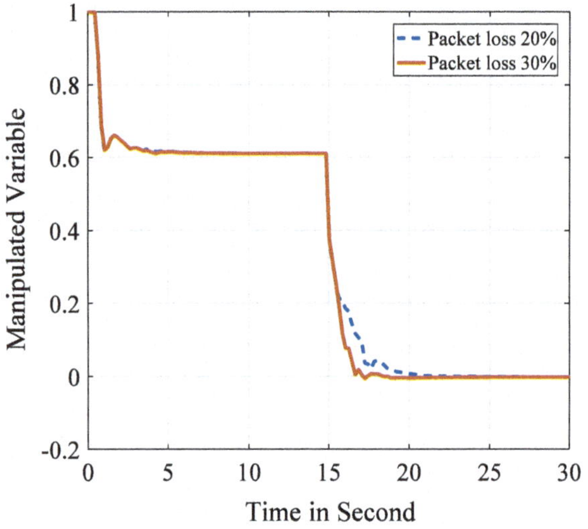

Fig. 4 Estimation of manipulated variable with optimal control

increase of packet loss probability and attack. This may lead to instability but the introduction of event-triggered optimal control mechanism maintained delay bound within acceptable limit.

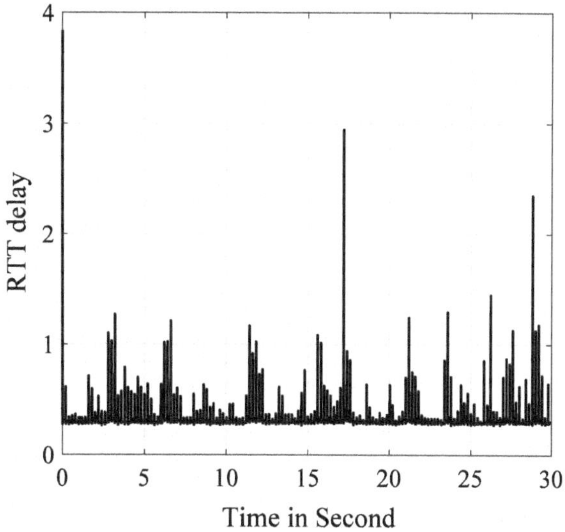

Fig. 5 Round-trip time delay with optimal control

5 Conclusion and Future Perspective of Work

The performance of industrial automation is of prime requirement to be maintained due to advancements in technologies. The attacker may introduce stochastic nature of dynamics in control system to degrade the stability and transient performance. The unwanted stochastic dynamics (denial-of-service, packet loss/dropout) are considered to evaluate the performance of industrial automation following event-triggered optimal control mechanism.

To detect abnormal behavior in industrial automation, attack sequence, and packet loss are introduced in the process. These stochastic uncertain dynamics use the Bernoulii-distribution to show network uncertainty and attacks on the process. The PI controller maintains the response at the desired level but as soon as packet loss and attacks are introduced by attacker into to system, it tends to be unstable. To counteract the stochastic nature of dynamics event-triggered optimal control mechanism is employed. The transient along with steady-state performance is kept at the reference level with the proposed methodology. This work may be extended to mitigate network imperfection with nonlinear systems and to avoid detection of system parameters to unable to design attack model.

References

1. Ajay, Mittal RK (2020) Incremental sheet forming technologies. CRC Press. https://doi.org/10.1201/9780429298905
2. Ajay, Singh H, Parveen, Almangour B (2023) Handbook of smart manufacturing: forecasting the future of Industry 4.0. In: Handbook of smart manufacturing: forecasting the future of Industry 4.0. CRC Press, pp 1–364. https://doi.org/10.1201/9781003333760
3. Ajay, Kumar A, Parveen, Goel R, Mittal RK (2023) Waste recovery and management. CRC Press. https://doi.org/10.1201/9781003359784
4. Amini A, Asif A, Mohammadi A, Azarbahram A (2021) Sampled-data dynamic event-triggering control for networked systems subject to DoS attacks. IEEE Trans Netw Sci Eng 8(3):1978–1990. https://doi.org/10.1109/TNSE.2021.3070804
5. Cao Y, Gao M, Zhang D (2016) Fault detection for networked control systems with uncertain time-varying sampling periods. In: Proceedings of the 28th Chinese control and decision conference, CCDC 2016, vol 1, pp 2956–2960. https://doi.org/10.1109/CCDC.2016.7531488
6. Caruntu CF, Rafaila RC (2017) Robust MPC for networked control systems with data-packet dropouts modeled as disturbances. In: 2017 21st International conference on system theory, control and computing, ICSTCC 2017, pp 152–157. https://doi.org/10.1109/ICSTCC.2017.8107026
7. Chemashkin FY, Zhilenkov AA (2019) Problems in network control systems and their solutions. In: Proceedings of the 2019 IEEE conference of Russian young researchers in electrical and electronic engineering, ElConRus 2019, pp 1165–1168. https://doi.org/10.1109/EIConRus.2019.8656665
8. Giri J, Al-Lohedan HA, Mohammad F, Soleiman AA, Chadge R, Mahatme C, Sunheriya N, Giri P, Mutyarapwar D, Dhapke S (2023) A comparative study on predication of appropriate mechanical ventilation mode through machine learning approach. Bioengineering 10(4):418. https://doi.org/10.3390/bioengineering10040418
9. Giri J, Ahmad N, Badawy AS (2024) Detection of Covid-19 based on convolutional neural networks using pre-processed chest X-ray images. AIP Adv 14(3). https://doi.org/10.1063/5.0200397
10. Gu Z, Park JH, Yue D, Wu ZG, Xie X (2021) Event-triggered security output feedback control for networked interconnected systems subject to cyber-attacks. IEEE Trans Syst Man Cybern: Syst 51(10):6197–6206. https://doi.org/10.1109/TSMC.2019.2960115
11. Hongjie L (2012) Robust H infinity control of networked control systems with randomly varying delay and missing measurements. In: Chinese control conference, CCC, pp 2753–2758
12. Hu L, Wang Z, Naeem W (2016) Security analysis of stochastic networked control systems under false data injection attacks. In: 2016 UKACC International conference on control, UKACC control 2016, pp 1–6. https://doi.org/10.1109/CONTROL.2016.7737519
13. Hu Z, Deng F, Xing M, Li J (2017) Modeling and control of Itô stochastic networked control systems with random packet dropouts subject to time-varying sampling. IEEE Trans Autom Control 62(8):4194–4201. https://doi.org/10.1109/TAC.2017.2676986
14. Hu J, Liu C, Song Y (2017) Switching control for networked control system with denial-of-service attacks. In: Chinese control conference, CCC, pp 7667–7672. https://doi.org/10.23919/ChiCC.2017.8028569
15. Huang C, Bai Y, Zhu Y (2010) PID controller design for a class of networked cascade control systems. In: Proceedings—2nd IEEE international conference on advanced computer control, ICACC 2010, vol 1, pp 43–47. https://doi.org/10.1109/ICACC.2010.5486782
16. Kajal A, Nandal SK (2020) A hybrid approach for cyber security: improved intrusion detection system using ANN-SVM. Indian J Comput Sci Eng 11(4):412–425. https://doi.org/10.21817/indjcse/2020/v11i4/201104300
17. Karthik G, Parthiban A, Sathish T, Giri J (2024) Experimental analysis and development of fuzzy model for the prediction of COP and mass flow rate in air-conditioning system. Interactions 245(1). https://doi.org/10.1007/s10751-024-01979-x

18. Kumar A, Mittal RK, Haleem A (2022) Advances in additive manufacturing: artificial intelligence, nature-inspired, and biomanufacturing, pp 1–500. https://doi.org/10.1016/C2020-0-03877-6
19. Kumar L, Ajay, Sharma RK, Praveen (2023) Smart manufacturing and Industry 4.0 state-of-the-art review. In: Handbook of smart manufacturing: forecasting the future of Industry 4.0, pp 1–28. https://doi.org/10.1201/9781003333760-1
20. Kumar A, Kumar P, Srivastava AK, Goyat V (2023) Modeling, characterization, and processing of smart materials. In: Modeling, characterization, and processing of smart materials, pp 1–370. https://doi.org/10.4018/978-1-6684-9224-6
21. Liu Y (2021) Secure control of networked switched systems with random DoS attacks via event-triggered approach. In: Chinese control conference, CCC, 2021-July(61803253), pp 1399–1404. https://doi.org/10.23919/CCC52363.2021.9550730
22. Liu T, Jiang S, Pan F (2016) Analysis and design of the time-delay compensation for networked control systems with random communication delay. In: Chinese control conference, CCC, 2016-August, pp 7234–7239. https://doi.org/10.1109/ChiCC.2016.7554502
23. Lu R, Xu Y, Zhang R (2016) A new design of model predictive tracking control for networked control system under random packet loss and uncertainties. IEEE Trans Industr Electron 63(11):6999–7007. https://doi.org/10.1109/TIE.2016.2585543
24. Lu W, Yin X, Fu Y, Gao Z (2020) Observer-based event-triggered predictive control for networked control systems under dos attacks. Sensors (Switzerland) 20(23):1. https://doi.org/10.3390/s20236866
25. Lucia W, Gheitasi K, Ghaderi M (2021) Setpoint attack detection in cyber-physical systems. IEEE Trans Autom Control 66(5):2332–2338. https://doi.org/10.1109/TAC.2020.3004326
26. Ma J, Liu K, Sun S, Yang G (2018) Design and simulation of large delay wireless network control system. In: Proceedings—8th international conference on instrumentation and measurement, computer, communication and control, IMCCC 2018, pp 217–220. https://doi.org/10.1109/IMCCC.2018.00053
27. Madhuri KS, Mungara J (2021) Reinforcement learning for intrusion detection and improving optimal route by cuckoo search in WSN. Indian J Comput Sci Eng 12(6):1706–1770. https://doi.org/10.21817/indjcse/2021/v12i6/211206024
28. Mahatme C, Giri J, Al-Lohedan HA, Mohammad F, Sunheriya N, Chadge R, Giri P, Mallik S, Dhayalan M (2024) Performance evaluation of a mechanical ventilation simulation model for diverse respiratory complications. AIP Adv 14(3). https://doi.org/10.1063/5.0189452
29. Mao Y, Jafarnejadsani H, Zhao P, Akyol E, Hovakimyan N (2020) Novel stealthy attack and defense strategies for networked control systems. IEEE Trans Autom Control 65(9):3847–3862. https://doi.org/10.1109/TAC.2020.2997363
30. Moridi A, Armaghan S, Sedigh AK, Choobkar S (2011) Design of switching control systems using control performance assessment index. In: Proceedings of the world congress on engineering 2011, WCE 2011, vol 2, pp 1339–1342
31. Pan Y, Yang GH (2017) Event-triggered fuzzy control for nonlinear networked control systems. Fuzzy Sets Syst 329(8):91–107. https://doi.org/10.1016/j.fss.2017.05.010
32. Pang ZH, Fan LZ, Dong Z, Han QL, Liu GP (2022) False data injection attacks against partial sensor measurements of networked control systems. IEEE Trans Circuits Syst II Express Briefs 69(1):149–153. https://doi.org/10.1109/TCSII.2021.3073724
33. Sargolzaei A (2021) A secure control design for networked control system with nonlinear dynamics under false-data-injection attacks. In: Proceedings of the American control conference, 2021-May, pp 2693–2699. https://doi.org/10.23919/ACC50511.2021.9482950
34. Sathish T, SaiKumar D, Patil S, Saravanan R, Giri J, Aly AA (2024) Exponential smoothing method against the gradient boosting machine learning algorithm-based model for materials forecasting to minimize inventory. AIP Adv 14(6). https://doi.org/10.1063/5.0208491
35. Solanki BS, Renu K, Srinivasan S (2019) Stability and security analysis with identification of attack on industrial networked control system: an overview. Internetw Indonesia J 11(2):3–8. https://www.internetworkingindonesia.org/Issues/Vol11-No2-2019/iij_vol11_no2_2019_solanki.pdf

36. Solanki BS, Kumawat R, Srinivasan S (2021) An impact of different uncertainties and attacks on the performance metrics and stability of industrial control system. In: Sharma H, Gupta MK, Tomar GS, Lipo W (eds) Lecture notes in networks and systems, vol 204, pp 557–574. https://doi.org/10.1007/978-981-16-1089-9_44
37. Solanki BS, Kumawat R, Srinivasan S (2021) Averting and mitigating the effects of uncertainties with optimal control in industrial networked control system. In: Proceedings—2021 IEEE international symposium on smart electronic systems, ISES 2021, pp 316–318. https://doi.org/10.1109/iSES52644.2021.00080
38. Solanki BS, Kumawat R, Srinivasan S (2021) Synthesize the effect of intrusion and imperfection on networked-connected control system with optimal control strategy. In: 2021 10th international conference on information and automation for sustainability, ICIAfS 2021, pp 105–110. https://doi.org/10.1109/ICIAfS52090.2021.9605956
39. Solanki BS, Kumawat R, Srinivasan S (2022) Optimal switching control design for industrial networked control system with uncertain exogenous dynamics. Mater Today: Proc. https://doi.org/10.1016/j.matpr.2022.11.202
40. Solanki BS, Kumawat R, Srinivasan S (2022) Mitigating effect of uncertain exogenous dynamics by parametric performance improvement with optimal control design. Int J Eng Trends Technol 70(5):346–356. https://doi.org/10.14445/22315381/IJETT-V70I5P238
41. Solanki BS, Kumawat R, Srinivasan S (2022) Optimized control function with estimation of system parameters against attack for networked control system. In: Intelligent computing techniques for smart energy systems. Lecture notes in electrical engineering, vol 862, pp 515–528. https://doi.org/10.1007/978-981-19-0252-9_47
42. Song Z, Zhou X (2013) Research and simulation of wireless sensor and actuator networked control system. In: 2013 25th Chinese control and decision conference, CCDC 2013, pp 3995–3998. https://doi.org/10.1109/CCDC.2013.6561648
43. Tang Y, Peng C, Yin S, Qiu J, Gao H, Kaynak O (2013) Robust model predictive control under saturations, pp 1–9
44. Vardhini KK, Sita Mahalakshmi T (2020) Implementation of swarm optimized artificial neural network for network intruder detection and attack classification. Indian J Comput Sci Eng 11(2):195–203. https://doi.org/10.21817/indjcse/2020/v11i2/201102166
45. Wang T, Gao H, Qiu J (2016) A combined adaptive neural network and nonlinear model predictive control for multirate networked industrial process control. IEEE Trans Neural Netw Learn Syst 27(2):416–425. https://doi.org/10.1109/TNNLS.2015.2411671
46. Wu G, Sun J, Chen J (2018) Optimal data injection attacks in cyber-physical systems. IEEE Trans Cybern 48(12):3302–3312. https://doi.org/10.1109/TCYB.2018.2846365
47. Xia Y, Chen J, Zhou L (2007) Networked control systems with different control inputs. In: Proceedings of the 26th Chinese control conference, CCC 2007, vol 6, pp 539–543. https://doi.org/10.1109/CHICC.2006.4347088
48. Xu H, Jagannathan S (2012) Stochastic optimal controller design for unknown networked control system under TCP. Proc Am Control Conf 1:6503–6508. https://doi.org/10.1109/acc.2012.6315009
49. Yang Y, Li Y, Yue D, Tian YC, Ding X (2021) Distributed secure consensus control with event-triggering for multiagent systems under DoS attacks. IEEE Trans Cybern 51(6):2916–2928. https://doi.org/10.1109/TCYB.2020.2979342
50. Zhang J, Peng C (2020) Guaranteed cost control of uncertain networked control systems with a hybrid communication scheme. IEEE Trans Syst Man Cybern: Syst 50(9):3126–3135. https://doi.org/10.1109/TSMC.2018.2833203
51. Zhang Q, Qu F (2019) Stabilization of networked control systems via switching and impulsive controllers. In: Proceedings of the 31st Chinese control and decision conference, CCDC 2019, pp 1852–1856. https://doi.org/10.1109/CCDC.2019.8832402
52. Zhang C, Yang F, Ye H (2012) Informative property of the data set in a single-input single-output (SISO) closed-loop system with a switching controller. Chin J Chem Eng 20(6):1128–1135. https://doi.org/10.1016/S1004-9541(12)60597-7

53. Zhang J, Peng C, Masroor S, Sun H, Chai L (2016) Stability analysis of networked control systems with denial-of-service attacks. In: 2016 UKACC international conference on control, UKACC Control 2016, pp 1–6. https://doi.org/10.1109/CONTROL.2016.7737622
54. Zhao N, Xing W, Shi P, Lim CP (2022) Event-triggered control for networked systems under denial of service attacks and applications. IEEE Trans Circuits Syst I Regul Pap 69(2):811–820. https://doi.org/10.1109/TCSI.2021.3116278
55. Zhen S, Hou Z, Yin C (2017) A novel data-driven predictive control for networked control systems with random packet dropouts. In: Proceedings of 2017 IEEE 6th data driven control and learning systems conference, DDCLS 2017, pp 335–340. https://doi.org/10.1109/DDCLS.2017.8068093
56. Zhu C, Chai F (2019) Triggered networked control system with random delay. In: Proceedings of 2019 IEEE 3rd information technology, networking, electronic and automation control conference, ITNEC 2019, (ITNEC), pp 1145–1149. https://doi.org/10.1109/ITNEC.2019.8729071

Wind Energy Converters and Controllers: Practical Aspects

Jayprakash Giri, Neeraj Kumar Mishra, and Ashwani Tapde

Abstract Wind energy is a fast-growing source of clean energy, and this chapter addresses its integration challenges, including grid connectivity, wind variability, and turbine placement. A complex controller, electrical converter, and generator are required when wind turbines are connected to the electrical grid in order to manage the variable power generation brought on by different wind speeds. It is essential to examine control strategies such as phase lock loops, pitch controllers, and passive stalls to maximize output and maintain steady voltage. Also discussed is the classification of Wind Energy Conversion Systems (WECS) into offshore and land-based setups, which highlight the advantages of offshore systems, such as their high-power output and confined footprint. Aside from that, it examines the physical and operational challenges that come with downwind rotors, including extreme cyclic loads on the blades and the impact of variable wind on rotor mechanics. Power stability can be enhanced by integrating WECS with grid systems using innovative technologies like HVDC and energy storage systems.

Keywords Converters and controllers · Wind Energy Conversion Systems (WECS) · Advanced control strategies · Grid stability · Energy production optimization · Converter technologies

1 Introduction

Although wind energy is the renewable energy source that is growing the quickest, there are still problems with wind unpredictability, grid connectivity, and turbine location. It takes complicated controllers, power converters, and generators to integrate wind turbines into the electrical system. Since the power produced by the Wind

J. Giri (✉) · N. K. Mishra
Electrical Engineering, NIMS University, Jaipur, Rajasthan, India
e-mail: jayprakash.giri@powergrid.in

A. Tapde
TIEIT, Bhopal, Madhya Pradesh, India

A. Kumar et al. (eds.), *Advances in Mechatronics Systems*, Emerging Trends in Mechatronics, https://doi.org/10.1007/978-981-97-8472-1_3

Energy Conversion System (WECS) (as shown in Fig. 1) is closely correlated with wind speed, control strategies must be implemented to maintain a steady voltage and the highest possible output. Recently, a lot of controversy has been there about wind energy conversion systems (WECS), by converting air motion into electrical power, it attempts to harness its mechanical power. Depending on where it is set up, this system falls under the category of either land-based or OWEC (offshore wind energy conversion). Even though OWEC was developed later than land-based wind energy technological advances, its considerable advantages—high power output, low land occupation, abundant wind energy, minimal wind shear—have drawn more scrutiny to it. Numerous challenging issues must be overcome for large-scale wind power plants to be designed and operated successfully. Both the angular velocity and force should remain constant if the system is intended to generate A.C. power. Regretfully, the wind speed on a large rotor is not continuous from top to bottom, nor is it uniform in direction or amplitude. As a result, the turbine blades experience extreme cyclic loads, which leads to fatigue issues. If a downwind rotor mechanism is employed, this issue is exacerbated since the support tower's shadow unloads the blade. Not to mention, this effect makes a loud popping sound that some people find annoying. Due to the fact that turbines generate electricity from wind flow, power regulation is essential in their design. Overestimation power can cause imbalanced loads and excessive vibrations, which can result in mechanical breakdowns.

Three controlling techniques are utilized to limit the generation of electricity under higher wind speeds: phases lock loop, grids alongside one another and pitch controllers, as well as active and passive stall and pitching controllers. The complete electrical circuit is shielded by a crow bar circuit. To extract the most energy possible from a renewable resource, several optimization approaches are applied, and the outcomes are modeled and simulated. Increasing requirements and challenges about leaving no carbon have an effect on have made inexperienced power techniques like wind power more famous. Grid integration of wind energy has a number of promise, but exceptional output power calls for the appropriate type of controllers. With improvements in generators, power electronics, and turbines, wind power generation is a technology that is expanding quickly. It can save operating expenses,

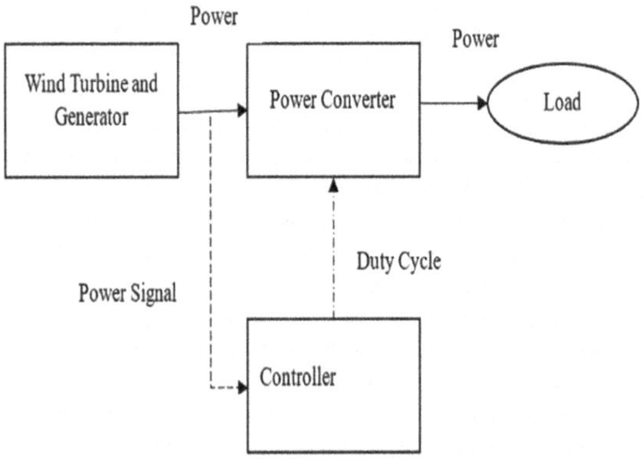

Fig. 1 Simple schematic representation of WECS

enhance voltage, and cut down on power losses. WECS, are a common tool for producing sustainable, clean, and green power. By 2020, WECS is estimated to have installed more than 280 GW globally, with a capacity of 1900 GW. However, the monitoring of wind power plants is challenging due to the recurrent nature of wind electricity [1]. A practical method for integrating the grid is through HVDC and ESS can enhance power stability and smoothing. Custom power devices and power electronic converters are useful tools for grid integration [2]. Low voltages, less than 690 Vrms line-to-line AC, are the main voltage range at which wind turbines produce power. New power electronics technologies are projected to achieve efficient and dependable power conversion at higher voltage levels (1–10 kV) to accommodate the increasing capacity of wind turbines and wind farms. Full-scale power-converter-based WTSs prefer multilevel converters because they can convert greater voltages and powers. One example of a multilevel converter is the marketed 3L-NPC/ANPC converter. Medium-voltage power conversion at a lower current density and reduced filter size is made possible by this converter, which achieves an additional voltage level and reduced dv/dt stress. Growing capacity for both offshore and onshore wind energy necessitates higher wind turbine reliability and maintainability. There is a growing use of multiphase motors in wind power production, particularly offshore wind power where the environment is unfavourable [3]. With them, you can operate at low voltages and high powers, and you can control them with fault-tolerant control. There are several advantages associated with transformer-less grid-connected operation at moderate voltages, including fault tolerance, increased reliability, a greater degree of control freedom, and design flexibility. Low-voltage and high-power generation devices at low-power levels are also included. Due to these benefits, multiphase motors are appropriate for big installed capacity offshore wind turbines and are especially useful for offshore wind generation in unfavourable environmental circumstances. WECS are expanding quickly and needs design and deployment changes. These networks of systems, often called cyber-physical systems (CPSs), are essential to the Internet of Energy, also known as IoE, a cloud-based system that connects renewable-energy (RE) sources to smart electrical grids and large-scale utility equipment like intelligent constructions and automobiles powered by electricity.

Less emphasis is being placed on generation systems in favour of demand-side management and smart buildings in research trends. The recommended releasable KE-based inertial controlling approach for DFIG, Wind power production aims to increase the frequency at which it operates by adjusting the two loop gains within each controller [4]. More KE is released when the DFIG controller runs the rotor at a faster speed. By improving the frequency nadir by greater than traditional gain systems, the suggested scheme ensures stable operation and keeps the rotor speed from exceeding the minimum speed limit. The effectiveness of the DFIG controller was examined in a range of wind scenarios. Energy storage is a critical element of sustainable development since renewable energies are revolutionizing the way we distribute and consume energy. Electric power systems employ a variety of storage technologies, such as thermal, mechanical in nature, electromagnetic waves, chemical, and electrochemical storage. Batteries used in electrochemical storage have a long lifespan under cyclic charge–discharge conditions, low self-discharge, and high efficiency [5]. Consumption of an electrolyser, fuel cells, and a pressurized gasoline tank, hydrogen energy storage (HES) converts electrical energy into chemical energy in the form of hydrogen Energy Storage methods involving mechanical power include flywheels, compressed air, hydroelectric pumping, and exceptional capacitors. Energy storage using superconducting magnets, and thermal storage using heat.

2 Power Quality and Voltage Regulation in Wind-Powered Systems

Power quality and voltage regulation are important components of integrating wind-powered systems into the electric grid. Wind strength, being inherently variable because of the fluctuating nature of wind speeds, poses a specific confrontation to maintaining uniform power quality and voltage ranges. These conflicts include flicker, transient disturbances, harmonic distortions, and voltage variations, all of which can impair the grid and wind-powered buildings' dependability and efficiency. To deal with these problems, advanced control strategies and technologies are deployed. Those include using static VAR compensators (SVCs), flexible AC transmission systems (data), and advanced inverter technologies for dynamic voltage law. These structures work through unexpectedly adjusting the voltage and reactive power flow, ensuring that voltage levels remain within suitable limits and that the power quality meets grid standards.

Although, wind generators geared up with variable-speed drives can contribute to voltage regulation and power quality improvement by adjusting their output in response to grid conditions. Power storage systems, which include batteries and supercapacitors, also are increasingly used alongside wind farms to buffer short-term fluctuations and make sure a constant power deliver. Ensuring appropriate strength high quality and voltage regulation in wind-powered systems is important for maximizing the penetration of energy created by the wind into the power grid retaining the grid's stability, and also ensuring the efficient operation of each wind generator and different related electrical devices. As wind power continues to develop as a significant element of global energy portfolios, ongoing research and development in these areas stay paramount.

For equipment and end-user architecture to function safely and efficiently, distribution systems must meet certain power quality requirements. As seen in Fig. 2, the aforementioned requirements address harmonics, power factor correction, and voltage as well as frequency regulation. Technical features and end-user operation of electrical loads present challenges since they can require a lot of power, require start-up, or produce harmonics. An offshore wind-powered petroleum and natural gas platform's reliability problems are examined in the investigation conducted in [6]. It highlights how load and wind-based power variations affect frequency and voltage variations in electrical systems including suggests strategies for intelligent load management and energy storage control. In [7], a multi-objective optimizing framework for building standalone microgrids (S-MG) with significant wind penetrating under erratic loads is provided. A framework facilitates decision-making about optimal component sizes, system prices, and harmonic distortion levels. The elitism non-dominated distinguishing genetic algorithm was able to resolve the issue of sorting. On Stewart Island, New Zealand, designing wind-powered environmentally friendly energy systems for unpredictable residential loads necessitates a thorough investigation of power quality issues, something the advised modelling technique effectively manages.

In the study referenced as [8], a three-phase self-excited induction generator paired with a nonlinear load converter is explored for enhancing power quality within wind energy systems. The generator faces performance issues, including harmonics, a low power factor, load imbalance, and neutral current problems, primarily due to voltage and frequency fluctuations triggered by load variations. To regulate frequency, a voltage source converter is utilized, and various control algorithms like LAF, LMS, VCO-less PLL, AVF, and NAVF have been assessed for their efficacy in generating reference currents and improving converter power quality characteristics. Additionally, the practicality of a small, self-sustaining wind-powered network incorporating energy storage is investigated in [9], where batteries are employed to store surplus energy. This storage ensures that in times when wind-generated energy falls short, the battery system can meet the load demands, maintaining proper voltage, frequency, and system parameters. By combining a solitary wind power system with a storage

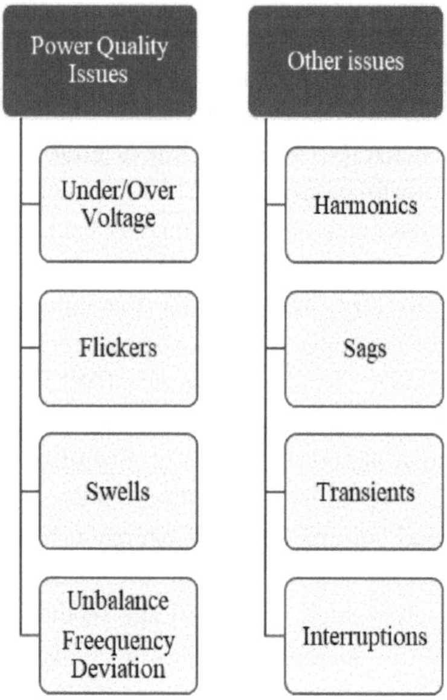

Fig. 2 Some common issues which affects the power quality

battery, experimental research has demonstrated that it is possible to create a system of renewable energy that emits zero emissions.

Despite the increasing presence of wind turbines and plug-in hybrid electric vehicles (PHEVs) in contemporary power grids, their contribution to primary frequency control in microgrids remains limited. The absence of cooperative control strategies that account for the limitations of both sources is noted in the existing literature. Therefore, the research presented in [10] employs small-signal analysis to determine the most effective frequency control strategy for combined PHEV and wind generator operation. The inclusion of renewable energy sources, like wind power, complicates the management of voltage and reactive power in power networks, leading to reactive power imbalances. Enhancing the system voltage/reactive power performance of the AVR and STATCOM has been proposed through a combined strategy that takes advantage of both particle swarm optimization and bacterial foraging optimization. The efficacy of a newly proposed controller is scrutinized using the H∞-loop shaping method, focusing on robustness and stability, and compared with traditional PI controllers within a wind-diesel system, as detailed in [11]. In order to avoid the need for a centrally controlled communication the system, it is suggested in [12] to use a frequency regulation technique to regulate wind-diesel electrical distribution networks that makes use of distributed energy generators (DEGs) in addition to independent distributed secondary loads (DSLs) with electrically powered thermally stored equipment. Frequency and voltage regulation in the wind-diesel mode are significantly influenced by factors such as total capacity, switching time, unit synchronization timing, and controller gain. The paper in [13] tackles the issue of wind power fluctuation and stability in IEEE 5 bus systems by proposing a Unified Power Flow Controller (UPFC) to improve WECS power quality. By connecting a UPFC to a grid with a wind generator, it significantly mitigates faults and enhances voltages at the point of common coupling (PCC), as well as real and reactive power outputs. Renewable energy sources, such as wind power, have a variable quality of electricity. With the aim of addressing this issue, the International Electrotechnical Commission's IEC-61400 standards are followed for assessing wind turbine performance and quality in terms of active and reactive power, voltage changes, harmonics, switching and flicker. To decrease the reactive power demands and serve as a primary source of electricity for the induction generator, a design that incorporates a static compensator and a storage system of energy has been projected to refine the quality of power. The Current source PWM inverters (CSPWMIs) are developed and evaluated for their characteristics, including voltage, load behaviour, harmonic distortion, load current and power factor. Simulink is used as a design for the CSPWMI, considering factors such as the voltage of the induction generator, the voltage of the build-up, the load voltage, etc. The energy resources that are renewable such as wind power, have a flexible quality of electricity. Simulink is used to design the CSPWMI, considering factors such as the voltage of the induction generator, the load voltage, the voltage of the build-up, etc. Additionally, [14] explores the use of a small-rating superconducting magnetic energy storage system in conjunction with a standalone wind-diesel system for power quality management, highlighting the system's enhanced voltage and frequency control through simulation and genetic algorithm tuning. Finally, [15] points to the emergence of electric springs and electric vehicles (EVs) as innovative components of power networks, embracing technological advancements and green energy solutions, with a focus on a wind-energy-conversion system integrated into a remote microgrid alongside a smooth band charge controller, showcasing the benefits of this approach compared to rapid transition controllers and marking a significant step towards widespread adoption of environmentally friendly technologies.

3 Wind Energy System Modelling

Wind energy, a renewable resource, stands out for its zero-pollutant emission and its vast availability across various geographical locations. The capability of a wind energy system to produce electricity significantly depends on the wind potential at its designated location, making the choice of site with optimal wind conditions a critical factor for the system's techno-economic viability. Vertical axis wind turbines use wind energy to generate electricity by having rotors that revolve around a vertical shaft with blades aligned vertically. They have benefits including minimal noise output and excellent turbulence tolerance, and they are usually smaller than horizontally oriented turbines. They are perfect for producing power in densely populated locations and under erratic, gusty wind conditions. Despite their lower energy efficiency, they are frequently seen in campuses, buildings, and wind power facilities. They can be used with an external storage battery or connected to the nearby power grid. On offshore platforms, vertically oriented wind turbines are often utilized for commercial uses. Because of its power, horizontal axis turbines have historically been the main focus of the wind power business. Wind turbines that have a vertical rotor shaft that don't require a wind point are known as vertical-axis wind turbines or VAWTs. They work well with renewable energy systems and modest systems for wind energy. They are available in a range of shapes and sizes, such as grid-connected wind inverters and vertical magnetic windmills with permanent magnetic generators. By virtue of their form, layout, and positioning, wind turbines harvest energy from the wind. Although horizontal turbines slow down the incoming wind, they still need to stop 100% of the wind in order to operate at 100% efficiency. The latest innovation for producing electricity that is intended for less extensive urban markets is called horizontal axis wind turbines or VAWTs. Because of their lesser elevation and closer position to the ground, they have a lower visual and environmental impact, which facilitates quicker regular consumption, inspection, and repair. The wind turbine converts mechanical energy into electrical energy using a VAWT generator. The wind turbine wings catch wind from two-way roadways, and the generator is placed in medians to divide traffic lanes. The gearbox sends the turbine's improved revolutions to the generator. The electrical power produced by the generator is kept in reserve in a battery that can be refuelled. The energy is used by an automated lighting system. The components of the LDR circuit are the light dependent resistor (LDR), a resistor, a transistor, battery, and LED. The LDR circuit automatically turns on the LED when the amount of light decreases. This method allows lighting control systems to automatically turn lights on or off. As its rotating shaft faces the direction of the wind, the most common type of wind turbines are known as horizontal axis wind turbines (HAWT). They can have single, double, or triple blades on their aerodynamically built blades, which spin due to the aerodynamic lift of force.

Bernoulli's relation, which claims that high air speed produces low-pressure regions and low-pressure domains, is what causes the pressure differential across the blades. When HAWT blades are attached to the rotor of a generator that produces electricity, the generator can harness stronger winds and operate more efficiently.

3.1 Rotor Huband Blade

Service loads and unavoidable variations in wind speed can cause weary loads on wind turbine rotor blades. This expands the possibility of cracks forming in the adhesive layers holding the rotor blade's fiber composite components together. As the quantity of stress cycles rises, cracks may develop and eventually the entire rotor blade may break. Contemporary wind turbines must operate at higher efficiency, which forces a continuous weight reduction. This results in longer rotor blades and a higher slimness ratio. Because of this, there is a greater chance of developing cracks in the adhesive connections due to increased stresses. The internal beam of the glass fiber-reinforced epoxy blades of the rotor transfers stresses into the cast iron rotor hub. A four-point ball bearing and an automated lubrication system pound the blade. The elongated cast hub ensures the best possible transfer of loads into the tower by transferring torque and blade forces through an additional main bearing. For automated yaw control, it has three hydraulic yaw brakes and six electrically yaw gears.

3.2 Generator

During wind turbine operation, mechanical energy is converted into electrical energy, which produces electricity. When the load is bigger, it takes more mechanical force to turn the rotor. Wind turbine generators come in a variety of sizes and power outputs. During a windy day, the blades of the turbine push against each other, creating rotational motion. It is important to optimize the blade design so that the wind can capture as much energy as possible. In a generator, the rotor is two-poled and connected to the grid at a fixed frequency and phase. Shape determines the shape of AC currents, which can take several forms. The frequency control of the generator is necessary to connect the power output to the grid. It is required to control the rotational speed in wind turbine applications. Using slip rings, generator brakes, offset coils, and power electronics, motor power is limited and motor life is prolonged.

The aerodynamic power output of a wind turbine, P_{Turbine}, is given by the equation:

$$P_{\text{Turbine}} = 21\rho AC_p(\lambda, \beta)v^3$$

where A is the area covered by the blades of the rotor (in square meters), C_p is the power coefficient, that differs with the tip-speed ratio (λ) and the angle of the blade pitch (β), and v is the speed of wind (in m/s), ρ is the air density, which is typically approximately 1.225 kg/m^3.

The tip-speed ratio, λ, is calculated as:

$$\lambda = v\omega mR$$

Here, ωm denotes the rotor's angular velocity (in radians per second), and R is the rotor radius (in meters).

The mechanical torque output from the wind turbine, Tm, is described by:

$$Tm = 1/2\rho AC_p(\lambda, \beta)\omega mv3$$

The power coefficient, C_p, a crucial factor in determining a turbine's efficiency, depends nonlinearly together on the tip-speed ratio and the blade pitch angle. It can be modeled generically as:

$$C_p = \frac{1}{2}\left(\frac{116}{\lambda_i} - 0.4\beta - 5\right)e^{-\left(\frac{21}{\lambda_i}\right)}$$

This equation considers the inherent characteristics and design parameters of the turbine to predict its performance.

To optimize the efficiency of wind energy extraction from the wind, WECS utilizes the concept of MPPT. In order to capture the maximum amount of power possible from the wind, the system adjusts the rotor speed according to the current wind conditions in real-time. Adapting dynamically to wind speeds and turbine designs is essential for maximizing wind turbine efficiency and output, showing the complicated interaction between wind speed and turbine design. This technique attempts to maximize the voltage and current outputs of the waveform by using the parameter that controls torque output wk.

4 Control Strategies for Wind Power System

Although adverse weather conditions are something that wind turbines are made to resist, excessive rotating torques or speeds are not. They use different control systems for high wind speeds, but they have a cut-out speed at which brakes must apply to stop them. These involve pitch control and stall control, both of which are optional. Depending on the operational region and the wind turbines controlling

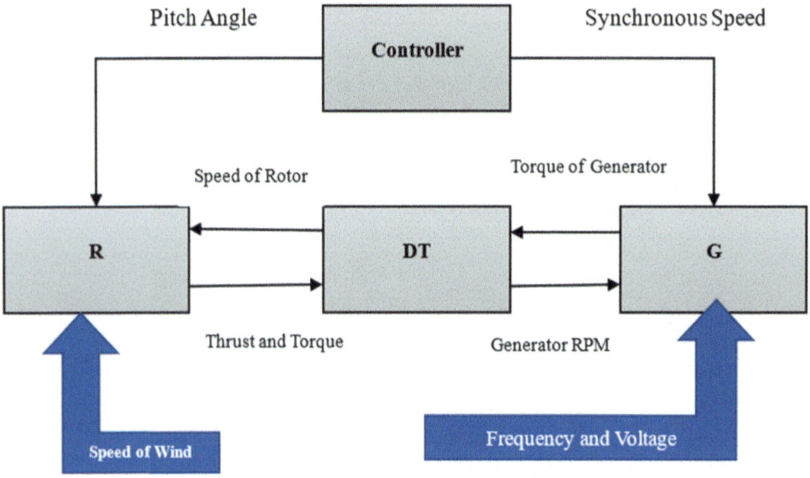

Fig. 3 Control methods of wind turbine.

objective, every control system has a different control approach. The discrete operational domains of any WT system are depicted in Fig. 3. The examination of any WT control approach requires the comprehension of several of these operational zones.

Where, G is Generator; R is Rotor and DT is Driving Train.

4.1 MPPT (Maximum Power Point Tracking) in Wind Turbines

Maximum Power Point Tracking (MPPT) algorithms are becoming more important for efficient energy transformation from energy sources that are renewable because of their capacity to manage nonlinear situations and their efficacy [16]. Provided the general patterns in MPPT schemes and clever strategies for wind-solar hybrid energy systems based on previous study surveys. The state of the art and effectiveness of new intelligent control systems will also be evaluated based on a number of criteria, including array dependence on others, speed of the wind, converging time, handling shadowing situations, and practical task execution. Employing a doubly fed induction generator (DFIG), as seen in Fig. 4, an RL filter, and a back-to-back converter to maximize power harvesting in the control of a WECS interfaced to a utility grid. The study carried out in [17] aimed to create a sensor-less regulator that could simultaneously adjust reactive power, generator speed, and DC-link voltage, therefore enhancing system dependability. The performance of the suggested controller is formally analyzed, and its practical viability is confirmed by numerical computations utilizing a 2 MW DFIG wind turbine (WT) under various operating scenarios.

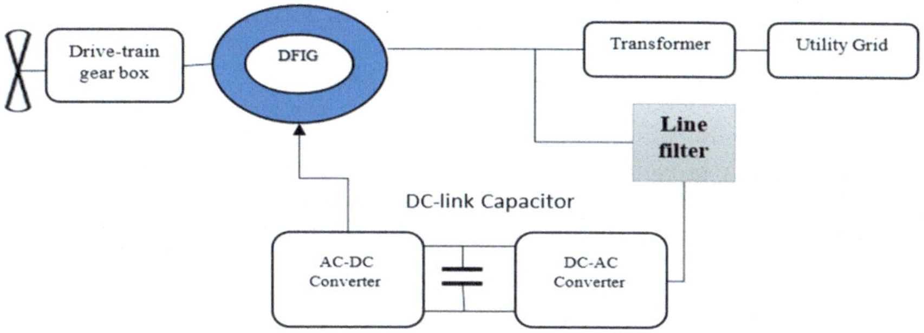

Fig. 4 Wind energy conversion system based on DFIG

Thermoelectric generators (TEGs) that harness the Seebeck phenomena are one green technology that can be utilized to boost energy efficiency by recovering a portion of waste heat as energy from electricity. The limitations of conventional MPPT algorithms include delayed convergence to the maximum power point (MPP), specific oscillations around the MPP, stuckness at the local maximum power point (LMPP), and inability to discover the global maximum power point (GMPP). In recent years, techniques utilizing metaheuristic MPPT algorithms have gained prominence as a means of addressing the shortcomings of traditional MPPT techniques. Confusion arises, though, when trying to choose the best approach amid the growing variety and sophistication of MPPT techniques. Reviewing in excess of sixty-two MPPT techniques that have been applied to TEGs in the previous six years with the possibility of being modified for TEGs and serving as a resource for researchers is the goal [18]. Solar power systems must always be operated at maximum capacity in order to be as efficient as possible. But as environmental factors like temperature, shade, and radiation change quickly, finding the maximum power that is available becomes a time-varying problem. MPPT methods are proposed in [19] to extract the greatest amount of power possible and track the ideal power point under these changing environmental conditions. Using MPPT to extract the greatest power is essential to creating PV systems that are efficient. These MPPT methods have a number of problems and restrictions, especially when partial shadowing is present due to uneven environmental conditions.

Renewable energy is a significant energy source that is constantly evolving. But little is known about it, particularly in Indonesia. ReVoLPoM for Renewable Energy System is a device that was developed. The study carried out in [20] aims to illustrate the process involved in developing this device. It's also known as the MPPT feature and it can track the highest power produced. The main topics of discussion are the ReVoLPoM algorithm's elements, modules, and pieces. The implementation of display utilities and data communication via a wireless network is intended to enhance system performance and facilitate user comprehension while obtaining crucial information from ReVoLPoM.

4.2 Grid Integration Controls: Voltage and Frequency Regulation

The frequency stability of power systems has been impacted by the deterioration of the inertia response brought on by the growing use of wind energy. Active power reserves can be employed to stabilize the frequency within a permitted range in order to resolve this. A variable speed wind turbine (VSWT) de-loading, capacitor storage in voltage source converter-high voltage direct current (VSC-HVDC) transmission, as depicted in Fig. 5, and frequency control coordination between offshore wind farms and VSC-HVDC transmission are the three methods. While HVDC can regulate frequency in the event of disturbances, it can also lead to power imbalances, which jeopardize the stability of the system [21]. In order to reduce load consumption in a fault-affected area, the study in [22] suggests an HVDC primary frequency regulation regulating voltage-dependent loads (PFR-VDL), which modifies grid voltage amplitude. Instead of drawing energy from generators, the PFR-VDL takes energy from voltage-dependent loads in the healthy region. Performance is examined for two potential HVDC connection scenarios in the study: embedded HVAC and asynchronous with a single line. Benefits and drawbacks are assessed using PSCAD EMTDC simulations, and a sizable networked IEEE 39 bus system is used for validation. The stability and dependability of the electrical grid are threatened by the rising usage of renewable energy sources and electric vehicles [22]. Vehicle-to-grid technology has the ability to reduce peak load and manage frequency at a profit. Using MATLAB/Simulink, a research conducted at the Canadian Centre for Housing Technology evaluated frequency controls based on single and multiple vehicles. By incorporating a coordinated control strategy for five combined electric vehicles, the model successfully decreased the frequency deviation inside a microgrid.

The shifting landscape of the energy sector, driven by deregulation, increasing consumption, and the rapid depletion of traditional fossil fuels, has significantly influenced the electrical grid's dynamics. This change has highlighted how crucial it is to include Renewable Energy Sources (RES) in the mix of energy sources. But there

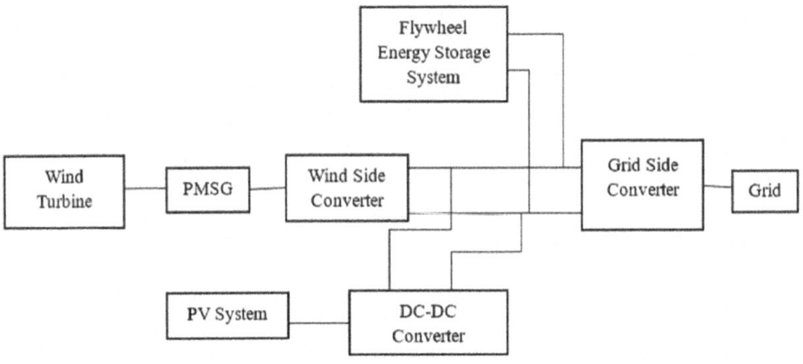

Fig. 5 VSC-HVDC transmission system integrated with flywheel energy storage system

are drawbacks to the comprehensive integration of these sources, namely preserving frequency stability and preventing voltage instability. Study [23] explores a range of configurations, including as microgrids, adaptable AC transmission networks, and storage of energy options, in order to assess the benefits and drawbacks of alternative control schemes. Within the context of distributed hybrid microgrids, the study explores the implementation of demand-response support (DRS) and virtual-inertia support (VIS) within an integrated resource planning framework [24]. This assessment utilizes real-time data from solar and wind sources modeled in MATLAB to craft a system design. A novel algorithm, the quasi-oppositional chaotic selfish-herd optimization (QCSHO), which blends quasi-oppositional-based learning with a chaotic linear search strategy, is introduced. The efficacy of QCSHO, demonstrated through a comparative analysis against seven other algorithms across five error metrics, highlights its potential superiority in system response optimization. The paper also addresses the complexities introduced by incorporating fluctuating energy sources into the grid, especially in terms of technical and economic implications for grid code standards in less robust power systems. The investigation in [25] focuses on the impact of grid fragility and isolation on the interconnection standards enforced by system operators on grid participants, offering a comparative analysis of generation asset regulations across different countries.

4.3 Pitch Controllers

Using ANFIS and MOA intelligent approaches, a new control strategy for WECS is introduced in [26]. An efficient optimization approach for testing and training the ANFIS controller is presented: MOA. The goal of developing a time-domain performance indicator is to simultaneously reduce settling time and system overshoot. On comparing with other strategies, the techniques of way it works outstanding reduction of settlement time and maximal overshoot—0.4497 s and 0.7989%, respectively—is validated. The proposed MOA-based ANFIS can handle wind speed variations, load demand variants, delay in response issues, and ambiguity among system parameters in addition to stabilizing WECS. The MOA-based ANFIS exhibits strong performance against variations in system characteristics, as confirmed by the uncertainty test. In order to mitigate the negative consequences of hydraulic pressure-driven unit delays, [27] attempts to enhance pitch control methods in wind turbines. An innovative approach is constructed by merging optimization techniques, delay-perturbation estimation purposes, and signal compensating approaches. Direct optimization for searches is used to develop a perfect delay-free pitch model and provide the most effective PI controller at the same time. PID-based pitch control systems are included in the application area of the technique. The blade-tower dynamics concepts of three turbines that generate electricity are analysed, and extensive simulation tests are conducted to demonstrate the effectiveness of the proposed techniques. The proposed pitch control methodology shows how delay perturbation estimation dependent compensating may successfully improve real-time pitch control efficiency employing a 4.8 MW benchmarking WECS. An improved pitch-angle control technique for a large-scale-WECS constructed around DFIG is provided by Barik et al. [28].

When wind speed exceeds the turbine's rated value, the control approach minimizes fatigue and adjusts the turbine's aerodynamic power. Appropriate system modelling and analysis are necessary for traditional control techniques, but they are not always available. The study suggests adaptive fuzzy-PID control, which has temporal variation, nonlinearity, multivariable, and time delay properties. The 1.5 MW DFIG WECS is subjected to the control approach. The controller's effectiveness is confirmed by the simulation results.

4.4 Voltage Controllers

Offshore wind presents challenges for maintaining frequency stability as renewable energy sources are integrated into electric power grids. A cutting-edge hybrid reloading control method allows wind energy conversion systems to regulate grid frequency using modular multilevel converters. Based on simulations of various wind speeds and load conditions, this technique bolsters frequency stability of AC grids by leveraging both grid-side and machine-side converters.

Also discussed in the paper is a feedback control-based proportional-integral (PI) controller for regulating wind energy systems with multiple rotors. According to the findings in [29], this approach reduces energy ripples, overshoots, steady-state errors, response times, and total harmonic distortion (THD) significantly by using pulse width modulation (PWM) for managing the double-fed induction generator via direct power control (DPC). Specifically, in changeable wind circumstances, a fractional PI (FPI) controller reduces active power fluctuations as well as overestimates by 95.42 and 90.86%, and current total harmonic distortion by 64.86 and 69.44%.

Furthermore, [30] offers a better wind turbine emulator (WTE) control method. This method generates the necessary power by simulating a wind turbine with nonlinear PI controllers. By adopting a drift-free Maximum Power Point Tracking (MPPT) methodology that makes use of the Perturb and Observe (P&O) technique, this method operates better than traditional approaches when it comes to control, emulation, and MPPT effectiveness. It emphasizes how crucial cutting-edge control techniques are to advancing the invention of renewable energy.

Overview of WECS technologies with respect to mechanical can be found in Table 1, which illustrates how pitch and voltage controllers are utilized. These papers examine different types of techniques, involving MOA-based ANFIS control, delay compensating optimization, and adaptive fuzzy-PID. As an outcome of this methodology, the settling time is shortened, the overshoot is reduced, and the present control is enhanced. Despite the fact that certain methods require extensive adjusting or modification for particular types of systems, there are significant improvements in THD, power ripples, and MPPT efficacy. Enhanced grid connections and increased wind turbine output are the results of the improved conversion of wind energy.

Table 1 Comparison of different controllers for power quality enhancement

Feature/ Study	[26]: ANFIS with MOA (pitch controllers)	[27]: Optimization and delay compensation (pitch controllers)	[28]: Adaptive fuzzy-PID (pitch controllers)	[31]: MMC with hybrid deloading (voltage controllers)	[29]: PI with DPC (voltage controllers)	[30]: Nonlinear PI and MPPT (voltage controllers)
Approach	ANFIS controlled by MOA	Optimization, delay-perturbation estimation, signal compensation	Adaptive fuzzy-PID	Hybrid deloading via PLL-free GFM control and DC-link voltage control	PI controller using DPC and PWM	Nonlinear PI controller with drift-free MPPT
Objective	Reduce settling time and system overshoot	Enhance real-time pitch control performance	Minimize fatigue, adjust aerodynamic power	Support grid frequency stability	Reduce energy ripples, overshoot, THD	Improve control, emulation, MPPT performance
Key results	Settling time: 0.4497 s, overshoot: 0.7989%	Effective delay compensation demonstrated on a 4.8 MW turbine	Tested on 1.5 MW DFIG WECS	Verified under various wind speeds and loading conditions	Significant reduction in THD and power ripples	Superior to traditional methods in control and MPPT efficiency
Advantages	Handles wind speed variation, load demand, and system uncertainty	Ideal for mitigating hydraulic pressure-driven unit delays	Suitable for systems with nonlinearity and time delay	Enhances energy support for AC grid frequency stability	Lowers THD and improves response time	Enhanced emulation and MPPT performance
Limitations	–	Requires accurate optimization and delay estimation	May need complex tuning for optimal performance	Requires sophisticated control strategies	Specific to systems using PWM and DPC	Focused on emulator systems, may need adaptation for real-world applications

5 Adaptive Control and Optimization

With the aim to increase their reliability and effectiveness, wind energy converters require adaptive control and optimization technologies. The arrangement of this sophisticated technology with these technologies enables the turbine to adapt instantly to changing wind conditions, ensuring maximum efficiency. Machine learning and AI algorithms applied to cutting-edge sensor technology. Advances have lowered operating stress and increased energy yield by improving turbine blade aerodynamic models. Intelligent control systems ensure a smooth grid integration as well. With the aim of attaining cleaner, more sustainable energy solutions, adaptive control and optimization are crucial for strengthening wind energy's competitiveness and sustainability. Wind Energy Conversion Systems (WECS) are controlled with a new robust control strategy based on a novel D-decomposition technique. A segmented plant provides better damping, lowers oscillations, accelerates settling, and reduces overshoot. On the basis of [32], it is clearly estimated that overshoot has been reduced by approximately 2.2329% and settling time by approximately 9.2626 s when compared to the traditional metaheuristic approaches. For the operation of wind turbines equipped with dual-fed induction generators, [33] proposes the employment of a Hidden Layer Neural Network combined with Recurrent Neural Networks (HLRNN). Within the framework of Indirect Vector Control (IVC) systems powered by PI controllers, the HLRNN approach effectively eradicates instability and ensures self-adaptation. Comparative analyses with existing neural network models and PI controller-based IVC systems, as per simulations conducted in MATLAB/Simulink, reveal that this method significantly lowers the response time to 1.8 ms, reduces static error to 0.16%, and decreases overshoot to 0.24%. Additionally, [34] presents an adaptive control strategy incorporating harmonic compensation for model reference in grid-connected converters. This method enhances the adaptive super-twisting sliding mode control by integrating additional actions to mitigate harmonic distortion of fundamental electrical grid elements, resulting in a reduced total harmonic distortion (THD) of 2% compared to the 2.5% observed without harmonic compensation. Furthermore, an adaptive control technique for mitigating frequency oscillations in single-area smart microgrid (μG) systems is detailed in [35], showcasing its efficacy in maintaining frequency stability and managing variable loads. Hybrid Jaya-Balloon optimizers (JBOs) are used for defining the process in it. Using effective frequency management, the controller can control variable loads of electric vehicles and high-power generators. The JBO stabilizes the system parameters and minimizes disturbances due to fluctuations in the system load. Based on comparisons with the alternative methods, it is possible to obtain a frequency variation of 0.0015 Hz which is within IEEE standards. According to MATLAB simulation results, the method offers significant stability against shocks and uncertainty in the parameters. Contemporary wind power systems strive for high MPPT, quick wind speed tracking, and excellent performance in variable wind conditions. Through the use of a sensor-less MPPT approach, performance and costs are improved and costs are reduced. To control the speed of grid-connected PMSG, Harris Hawks uses its algorithm-based

PI controller (HHA-PIC). By comparison with traditional CPIC and AFLC systems, the HHA-PIC increases the efficiency of the system by 0.81 and 8.48%. The goal of this research was to create an MPPT approach for small-scale wind turbines (WTs) that addresses the drawbacks of conventional MPPT techniques.

By attaining quick speed and reduced oscillations, the method improves system performance. It monitors the MPP and keeps an eye on the operational rotor speed, using the rotation speed to determine the ME's slope. The method's ease of use, stability, and adaptability help integrate WTs into contemporary electrical networks on a big scale [36]. With a settling time of only 0.09 s when inertia is not considered, the studied MPPT method increases wind system efficiency and quickly tracks the MPP; when inertia is considered, the settling time is closer to 7.89 s. Table 2 offers a comprehensive overview of recent advancements in control and optimization strategies for wind energy converters, highlighting a variety of approaches from robust control and neural networks to adaptive control and innovative optimization algorithms.

Each study in Table 2, focuses on addressing specific challenges in wind energy systems, such as improving damping performance, enhancing system stability, compensating for harmonics, managing variable loads, and maximizing power output. The key features, results, and benefits are detailed, including quantitative metrics like settling time, overshoot, and other performance indicators. These studies collectively demonstrate significant improvements in wind energy converter performance, showcasing the potential for these technologies to enhance the efficiency, reliability, and integration of wind power into the energy grid.

6 Conclusion

Wind energy conversion systems are improved by innovative control strategies and power electronics. In order to mitigate the challenges posed by the variable nature of wind energy, robust controllers and advanced converters can seamlessly integrate wind energy into the power grid. In the study, adaptive control, D-decomposition, and self-adaptation outperformed traditional control methods for harmonic compensation. Increasing the reliability and efficiency of renewable energy sources boosts damping performance, reduces overshoot and settling times, and enhances MPPT and grid stability. In addition, it is necessary for wind energy systems to continuously innovate their control strategies in order to remain efficient as they develop. In order to harness wind energy more effectively and sustainably, interdisciplinary collaborations between engineers, IT specialists, and policy makers are necessary. Research such as this paves the way for future advancements in wind energy technology.

Table 2 Control strategies for enhanced performance in wind energy conversion systems

Study	Approach	Key features	Results	Benefits	Settling time	Overshoot	Other performance metrics
[32]	D-decomposition	Robust control, segments plants for best damping performance	Outperforms metaheuristics in damping performance	Less overshoot, minimal oscillation	9.2626 s	2.2329%	–
[33]	Hidden Layer Neural Network with Recurrent Neural Networks (HLRNN)	Improves stability and self-adaptation in dual-fed induction generator systems	Better than PI controller and some NN initiatives in response time and static error	Reduced response time, minimized static error	1.8 ms	0.24%	Static error: 0.16%
[34]	Harmonic compensation approach	Adaptive control for grid-connected converters, addresses harmonic contents	Lower THD compared to without compensation	Effective harmonic compensation	–	–	THD: 2%
[35]	Hybrid Jaya-Balloon Optimizer (JBO)	Manages variable loads, minimizes parameter fluctuations	Achieves frequency variation within IEEE standards	Frequency stability, handles shocks and uncertainty	–	–	Frequency variation: 0.0015 Hz
[37]	HHA-PI controller	Sensorless MPPT for grid-connected PMSG, estimates actual wind speeds	Increases system efficiency compared to traditional CPIC and AFLC	High MPPT, quick wind speed tracking	–	–	System efficiency increase: 0.81% compared to CPIC, 8.48% compared to AFLC
[36, 38–41]	MPPT based on SC theory	Efficient MPPT for small-scale WTs, quick speed and reduced oscillations	Quick MPP tracking, increases wind system efficiency	Stability, adaptability for large scale integration	0.09 s (without inertia), 7.89 s (with inertia)	–	Cell efficiency, output power improvement

References

1. Mahela OP, Shaik AG (2016) Comprehensive overview of grid interfaced wind energy generation systems. Renew Sustain Energy Rev 57:260–281. https://doi.org/10.1016/j.rser.2015.12.086
2. Laws ND, Epps BP (2016) Hydrokinetic energy conversion: technology, research, and outlook. Renew Sustain Energy Rev 57:1245–1259. https://doi.org/10.1016/j.rser.2015.12.087
3. Khaligh A, Onar OC (2017) Energy harvesting: solar, wind, and ocean energy conversion systems. CRC Press
4. Blaabjerg F, Ma K (2017) Wind energy systems. Proc IEEE 105(11):2116–2131. https://doi.org/10.1109/JPROC.2017.2720344
5. Peng X, Liu Z, Jiang D (2021) A review of multiphase energy conversion in wind power generation. Renew Sustain Energy Rev 147:111172. https://doi.org/10.1016/j.rser.2021.111172
6. Moness M, Moustafa AM (2015) A survey of cyber-physical advances and challenges of wind energy conversion systems: prospects for internet of energy. IEEE Internet Things J 3(2):134–145. https://doi.org/10.1109/JIOT.2015.2396471
7. Lee J, Muljadi E, Srensen P, Kang YC (2015) Releasable kinetic energy-based inertial control of a DFIG wind power plant. IEEE Trans Sustain Energy 7(1):279–288. https://doi.org/10.1109/TSTE.2015.2482351
8. Amrouche SO, Rekioua D, Rekioua T, Bacha S (2016) Overview of energy storage in renewable energy systems. Int J Hydrogen Energy 41(45):20914–20927. https://doi.org/10.1016/j.ijhydene.2016.08.100
9. Yuce MI, Muratoglu A (2015) Hydrokinetic energy conversion systems: a technology status review. Renew Sustain Energy Rev 43:72–82. https://doi.org/10.1016/j.rser.2014.11.033
10. Li Y, Xu Z, Wong KP (2016) Advanced control strategies of PMSG-based wind turbines for system inertia support. IEEE Trans Power Syst 32(4):3027–3037. https://doi.org/10.1109/TPWRS.2016.2522823
11. Alves E, Sanchez S, Brandao D, Tedeschi E (2019) Smart load management with energy storage for power quality enhancement in wind-powered oil and gas applications. Energies 12(15):2985. https://doi.org/10.3390/en12152985
12. Mohseni S, Brent AC, Burmester D (2020) Power quality considerations in the planning phase of stand-alone wind-powered micro-grids. In: 2020 19th international conference on harmonics and quality of power (ICHQP). IEEE, pp 1–6. https://doi.org/10.1109/ICHQP49022.2020.9249180
13. Giri AK, Arya SR, Maurya R, Ray P (2020) Performance of control algorithms in wind-based distributed generation system with power quality features: a review. In: Microgrid: operation, control, monitoring and protection, pp 51–109. https://doi.org/10.1007/978-981-13-9733-2_4
14. Wagle R, Sharma P, Sharma C, Gjengedal T, Pradhan C (2021) Bio-inspired hybrid BFOA-PSO algorithm-based reactive power controller in a standalone wind-diesel power system. Int Trans Electr Energy Syst 31(3):e12778. https://doi.org/10.1002/2050-7038.1277
15. Janssen NT, Wies RW, Peterson RA (2016) Frequency regulation by distributed secondary loads on islanded wind-powered microgrids. IEEE Trans Sustain Energy 7(3):1028–1035. https://doi.org/10.1109/TSTE.2016.2542020
16. Deve VS, Keerthivasan K, Gheetha K (2020) Power quality improvement for a wind farm connected grid incorporating UPFC controller
17. Kinge AP (2019) Review of Statcom in guiding the grid connected wind energy for power quality enhancement. Int J Innov Eng Res Technol 4(6):1–4
18. Abass AA, Mufti MU-D (2019) Sim power-based analysis and design of a hybrid wind–diesel-superconducting magnetic energy storage system for simultaneous frequency and voltage control. Wind Eng 43(6):596–608. https://doi.org/10.1177/0309524X19869834
19. Mohanty S, Pati S, Kar SK, Flah A, El-Bayeh CZ, Veerendra AS (2023) Development of wind-powered smart transition electric vehicle charging station. IET Electr Syst Transp

20. Chandra S, Gaur P, Srishti (2018) Maximum power point tracking approaches for wind–solar hybrid renewable energy system—a review. In: Advances in energy and power systems: select proceedings of ICAEDC, pp 3–12
21. Abouloifa A, Noussi K, Elbouchikhi E, Katir H, Lachkar I, El Aroudi A (2022) High-gain observer-based advanced nonlinear control of a grid-connected wind energy conversion system with sensorless maximum power point tracking. Machines 10(11):1074. https://doi.org/10.3390/machines10111074
22. Mamur H, Üstüner MA, Bhuiyan MRA (2022) Future perspective and current situation of maximum power point tracking methods in thermoelectric generators. Sustain Energy Technol Assess 50:101824. https://doi.org/10.1016/j.seta.2022.101824
23. Worku MY, Hassan MA, Maraaba LS, Shafiullah M, Elkadeem MR, Hossain MI, Abido MA (2023) A comprehensive review of recent maximum power point tracking techniques for photovoltaic systems under partial shading. Sustainability 15(14):11132. https://doi.org/10.3390/su151411132
24. Yue X, Zhang J, Meng F, Liu J, Chen Q, Geng D (2023) Multi-timescale lookup table based maximum power point tracking of an inverse-pendulum wave energy converter: power assessments and sensitivity study. Energies 16(17):6195. https://doi.org/10.3390/en16176195
25. Renaldo T, Djajadi A (2016) Developing a maximum power point tracking system for renewable energy. PhD diss., Swiss German University
26. Jallad J, Mekhilef S, Mokhlis H (2017) Frequency regulation strategies in grid integrated offshore wind turbines via VSC-HVDC technology: a review. Energies 10(9):1244. https://doi.org/10.3390/en10091244
27. Lagasse M, De Carne G, Liserre M, Biskoping M (2020) Primary frequency regulation using HVDC terminals controlling voltage dependent loads. IEEE Trans Power Deliv 36(2):710–720. https://doi.org/10.1109/TPWRD.2020.2973888
28. Barik AK, Das DC, Latif A, Hussain SMS, Ustun TS (2021) Optimal voltage–frequency regulation in distributed sustainable energy-based hybrid microgrids with integrated resource planning. Energies 14(10):2735. https://doi.org/10.3390/en14102735
29. Imam AA, Sreerama Kumar R, Al-Turki YA (2020) Modeling and simulation of a PI controlled shunt active power filter for power quality enhancement based on PQ theory. Electronics 9(4):637. https://doi.org/10.3390/electronics9040637
30. Elsisi M, Tran MQ, Mahmoud K, Lehtonen M, Darwish MM (2021) Robust design of ANFIS-based blade pitch controller for wind energy conversion systems against wind speed fluctuations. IEEE Access 9:37894–37904. https://doi.org/10.1109/ACCESS.2021.3061000
31. Etxegarai A, Eguia P, Torres E, Iturregi A, Valverde V (2015) Review of grid connection requirements for generation assets in weak power grids. Renew Sustain Energy Rev 41:1501–1514. https://doi.org/10.1016/j.rser.2014.08.058
32. Ali MN, Soliman M, Ebrahim MA, Elsisi M (2023) D-decomposition-based multi-objective robust resilient control for blade pitch of wind energy conversion system. Int J Electr Power Energy Syst 146:108781. https://doi.org/10.1016/j.ijepes.2021.107820
33. Chetouani E, Errami Y, Obbadi A, Sahnoun S (2023) Self-adapting PI controller for grid-connected DFIG wind turbines based on recurrent neural network optimization control under unbalanced grid faults. Electr Power Syst Res 214:108829. https://doi.org/10.1016/j.epsr.2022.108829
34. Hollweg GV, Evald PJ, Mattos E, Borin LC, Tambara RV, Gründling HA, Su W (2023) A direct adaptive controller with harmonic compensation for grid-connected converters. IEEE Trans Industr Electron. https://doi.org/10.1109/TIE.2023.3159745
35. Ewais AM, Elnoby AM, Mohamed TH, Mahmoud MM, Qudaih Y, Hassan AM (2023) Adaptive frequency control in smart microgrid using controlled loads supported by real-time implementation. PLoS ONE 18(4):e0283561. https://doi.org/10.1371/journal.pone.0283561
36. Chadge R, Giri J, Amir M, Bakhsh FI, Sunheriya N, Mahatme C, Sathish T (2024) Investigation of electrical configuration topologies for enhancing the performance of solar photovoltaic water pumping system. Electr Power Compon Syst 1–16. https://doi.org/10.1080/15325008.2024.2319312

37. Aldin NAN, Abdellatif WS, Elbarbary ZS, Omar AI, Mahmoud MM (2023) Robust speed controller for PMSG wind system based on Harris Hawks optimization via wind speed estimation: a real case study. IEEE Access 11:5929–5943
38. Sathish T, Giri J, Saravanan R, Ali MS, Muthukumaran N (2024) Solar FPC performance enrichment with Al_2O_3/SiO_2 nanofluids and hybrid nanofluid. Case Stud Therm Eng 60:104718. https://doi.org/10.1016/j.csite.2024.104718
39. Sungeetha D, Giri J, Makki E, Panchal H, Gomathi S, Deepa P, Kumar A, Praveenkumar TR (2024) An inimitable Elman network based fire hawk controller and skill optimized power tracker with ultra gain converter for improving the performance of PV tied EV systems. Case Stud Therm Eng 56:104183. https://doi.org/10.1016/j.csite.2024.104183
40. Sunheriya N, Mahakalkar SG, Chadge RB, Lande P, Giri JP, Bagde A (2022). Experimental analysis of twin blade contra-rotating wind turbine. https://doi.org/10.1109/icetems56252.2022.10093565
41. Chadge RB, Sunheriya N, Giri JP, Shrivastava N (2022) Experimental investigation of different electrical configurations and topologies for Photovoltaic system. Mater Today: Proc 57:316–320. https://doi.org/10.1016/j.matpr.2022.03.097

Dual-Band Planar Antenna Design Using Characteristics Mode Analysis for 5G Sub-6 Band and WLAN Application

Jyoti C. Kolte, Payal Bansal, and Ashwini Kumar

Abstract A dual-band planar antenna is presented for 5G sub-6 band and WLAN applications. The antenna is developed using characteristic mode analysis (CMA). It operates at 3.5 GHz for the 5G sub-6 band and 5.8 GHz for WLAN. The initial phase involved designing conventional circular patch antennas and subjecting them to CMA for a single band. To enhance the bandwidth at 3.5 GHz and enable dual-mode functionality, the circular shape was transformed into an elliptical one. This modification contributed to the antenna's ability to support both frequencies simultaneously. The achievement of dual-band characteristics is facilitated by incorporating an external circular ring within the elliptical structure. The external ring is specifically responsible for generating a lower mode at 3.5 GHz. A comprehensive full-wave analysis was conducted using the Finite Integration Technique (FIT) numerical approach, along with a coaxial feed, to thoroughly assess the performance of the proposed optimal design. The proposed antenna demonstrates desirable attributes for dual-band operation, making it well-suited for 5G and WLAN applications.

Keywords Dual-band · CMA · Sub-6 band · WLAN · FIT · WiMAX

1 Introduction

Multi-band technology is widely used in recent 5G mobile wireless communication because using this technology it is very convenient to switch the operation of one antenna band to another without changing the operation of the previous band. This technology provides advantages like improved selectivity, throughput, and also

J. C. Kolte (✉) · P. Bansal · A. Kumar
Poornima University, Jaipur, India
e-mail: Jyoti180181@gmail.com

P. Bansal
e-mail: Payal.Bansal@poornima.org

A. Kumar
e-mail: ashwini.kumar@poornima.edu.in

A. Kumar et al. (eds.), *Advances in Mechatronics Systems*, Emerging Trends in Mechatronics, https://doi.org/10.1007/978-981-97-8472-1_4

reduced complexity and power consumption. One of the key elements of wireless communications is the environment in which electromagnetic waves spread. Due to the reflection and diffraction effects caused by the objects that make them up, they are most frequently of the multipath type [1–3]. With the growth of numerous wireless communications standards, it is ideal to combine as many standards as possible into a single wireless device, including the Global Positioning System (GPS), Worldwide Interoperability for Microwave Access (WiMAX), and Wireless Area Network (WLAN) standards [4]. Researchers from all over the world are developing various methods to make antennas as small as feasible without losing or compromising the desired properties of current electronics [5–8]. Researchers from all over the world are working on various methods to reduce the size of antennas without compromising the desired qualities of current gadgets [9–14]. The applications of multi-band antenna are IoT, self-driving cars, machine-to-machine communication, biomedical implants, wireless sensor networks smart cities, etc. [15–19]. Sub-6 GHz range and MM-Wave bands make up the majority of the spectrum that the Federal Frequency Commission (FCC) has allotted for 5G applications. To increase the deployment process of 5G base stations with existing base stations, The sub-6 band has been allocated for the benefits stated in [20–27].

Due to its non-planar construction and versatility in placement on uneven surfaces, patch antennas are widely used in wireless systems. These antennas are lightweight, inexpensive, flexible, straightforward, and small, in addition to being straightforward [10].

Comparative Literature review has been shown in Table 1 which proves that FR4 substrate is widely used due to its ease of fabrication, whereas Microstrip patch antenna is widely used in 5G sub-6 band and WLAN frequency Range.

In the literature, a variety of methods are employed to create multi-band antennas. The DGS (Defected Ground Structure) and slots on the patch design methods are used frequently to create multi-band antennas. The patch also has multi-band capabilities in its loaded resonator slot. We can create multi-band antennas utilizing various optimization techniques, including the genetic algorithm. These antennas have a lot of advantages, but other performance aspects still need to be improved, such as size, circuit integration complexity, and radiation quality. The proposed novel antenna has been designed on an FR4 substrate, which is less complex and eases fabrication.

These antennas have numerous advantages, yet certain performance aspects still require enhancement, particularly in terms of size, circuit integration complexity, and radiation quality. The proposed novel antenna was designed using characteristic mode analysis, enabling it to exhibit dual-band characteristics. Notably, this design approach has significantly reduced complexity and Streamlined the fabrication process, addressing some of the key challenges associated with traditional designs. While the antenna offers dual-band functionality and improved fabrication ease, it aims to address the broader challenges within the field. Specifically, efforts are ongoing to minimize the antenna's overall size, simplify circuit integration, and optimize radiation quality, thereby ensuring superior performance across the 5G sub-6 band and WLAN applications [28]. The development of such antennas represents a step forward in the quest for more efficient and integrated communication solutions.

Table 1 Comparative analysis of different antenna design 5G sub-6 band and WLAN frequency range

References	Type of antenna	Dimension of antenna	Material for design	Operating frequency (GHz)	Gain (dB)
[5]	Dual band MSA	40 * 40 mm^2	FR4	52 MHz 157 MHz	–
[6]	CPW-fed dual band	24.5 * 20 mm^2	FR4	3.24–8.29 9.12–11.25	4.33 dBi 4.80 dBi
[7]	MSA patch	27.5 * 27.5 mm^2	FR4	2492.08 MHz	2.83 dB
[19]	MSA patch	33.33 × 33.33 × 0.16 cm^3	FR4	3.6 GHz	5.4 dBi
[20]	Dual polarized dipole antenna	27.2 * 27.2 * 17 mm^3	–	5.2 GHz	26.8 dB
[21]	MSA patch	75 * 75 * 14.8 mm^3	FR4	2.4	7 dB
[22]	Patch antenna design with two gap cuttings	12.8 * 19.8 cm^2	Rogers Ultralam1217	1.5 GHz	8.348 dB
[23]	Dual bands planar array	150 * 180 mm^2	Duroid RT5870	5.4 GHz 15 GHz	7.6 dBi 7 dBi
[24]	Dual-band monopole	70 * 50 * 0:51 mm^3	RO-5880	2.6 and 3.5 GHz	15 dBi
[25]	MSA patch	140 * 70 * 1 mm^3	FR4	4.25 GHz	–
[26]	Microstrip antenna	60 * 96 * 0.8 mm^3	FR4	3.3–3.7 GHz 5.1–5.9 GHz	13 dB gain
[27]	Two trapezoidal-shaped patches	17 × 42 mm^2	FR4	7.2 and 8.5 GHz	>5 dB and about 1.7 dB

The proposed antenna design work is summarized in the following sections. In Sect. 2, the modification and optimization work has been described using the CMA technique without feed excitation. For verification of the CMA methodology, a full-wave simulation of the proposed design is performed in Sect. 2.1 using a coaxial feed line. Finally, in Sect. 2.2, the proposed work is summarized

2 Antenna Design Principle and Analysis

The proposed antenna schematic shown in Fig. 1, is designed using FR4 material having dielectric constant 4.3 and loss tangent of 0. 025. This material is lightweight and low cost. It consists using one elliptical patch which has $R_U = 6.7$mm, and $R_V = 5.7$mm, and outer ring antenna has a radius $R = 10$ mm, it is connected from elliptical antenna by strip line with $d = 2.32$ mm and width $k = 3.12$ mm. The complete antenna volume is 60 * 60 * 1.6 mm^3.

The dimensions of proposed antenna design are mentioned in Table 2. Length if substrate $L = 60$ mm, width of substrate $W = 60$ mm, elliptical patch dimension are $R_U = 6.7$ mm, and $R_V = 5.7$ mm, outer ring radius $R = 10$ mm, strip line with $d = 2.32$ mm and width $k = 3.12$ mm.

2.1 A Characteristics Mode Analysis (CMA) of Propose Antenna

A strong tool for designing and conducting microwave circuits using a methodical approach like analysis, synthesis, and optimization is the characteristics mode theory. Garbacz, Harrington, and Mautz were the ones to develop this technique. The integral surface equation and the MoM (Method of Moments) are used to quantitatively

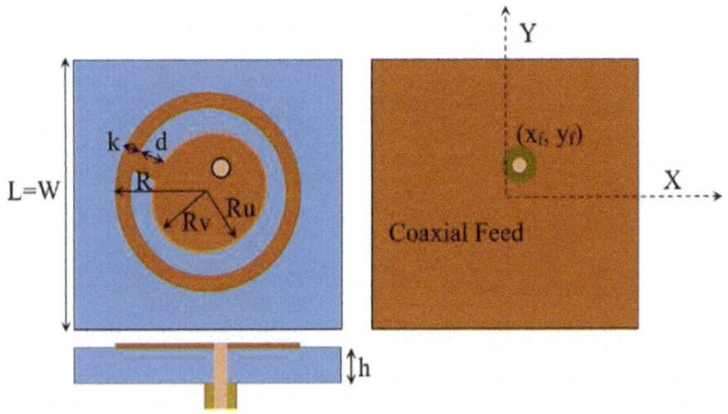

Fig. 1 Design schematic of propose antenna

Table 2 Proposed antenna parameters

Parameters	Values (mm)	Parameters	Values (mm)
L	60	W	60
R_u	6.7	R_v	5.7
R	10	d	2.32
k	3.12	h	1.6
xf	1.83	yf	1.83

compute the characteristic modes of conducting structures in the absence of stimulation [17, 18]. This is a systematic design process that offers a physical knowledge of the conducting structures in contrast to other existing methods like trial-and-error or automatic optimization, which hide the physical behavior of circuits. On the surface of the perfect electric conductor (PEC) structure, the characteristic mode introduces a large collection of orthogonal real current modes. These mode currents, which are characteristics of the conducting electromagnetic material itself, may be derived using the eigenvalue equation [19].

$$[X]\,, J_{n=}\lambda_n[R], J_n \ [2] \tag{1}$$

where, J_n and λ_n are the characteristics current and Eigenvalues of n modes respectively. Impedance matrix (Z) of conducting body is given in the equation

$$Z = R + jX \ [2] \tag{2}$$

The magnitudes of Eigenvalue provide resonance and radiating characteristics of the conducting body with dissimilar current modes. The Eigenvalues λ_n vary from $-\infty$ to $+\infty$, when the mode is in radiating condition then its corresponding Eigenvalue is zero ($\lambda_n = 0$). In the case of storing magnetic energy, it means positive ($\lambda_n > 0$) modes, and for storing electric energy negative ($\lambda_n < 0$) modes. The Model Significant (MS) and Characteristics (CA) are specified as [9, 18]:

A mode resonates, when $\lambda_n = 0$ According to the characteristic mode analysis, the FR4 substrate is used to create the circular patch antenna, using FR4 material having a dielectric constant of 4.3 and loss tangent of 0.025. Antennas (a, b, and c) in Fig. 2 are shown step-by-step dual band antenna designing using the CMA method. These are simulated up to four characteristic modes.

Figure 3 shows the modal significant plot of antenna-1 structure up to four modes. In this plot, mode-1 and mode-2 are resonant at 5.8 GHz frequency at $R_u = R_v = 5.8$ mm, because at this point the modes show $\lambda_1\ \lambda_2 = 0$, which means MSn values are approx. unity (MSn = 1) and both modes are overlapped. Other higher order modes

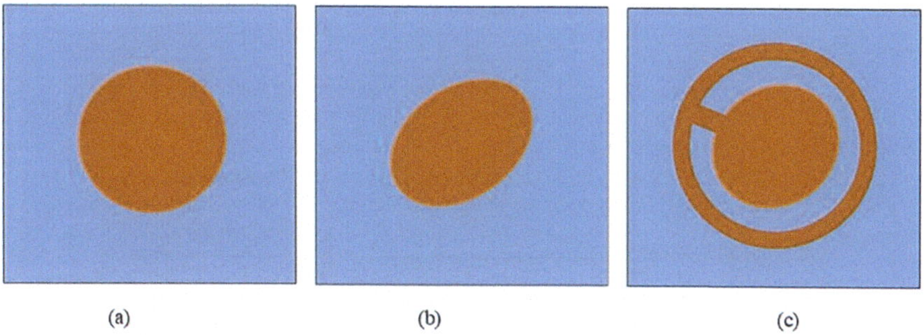

Fig. 2 CMA analysis of antenna, **a** antenna-1; **b** antenna-2; **c** antenna-3

have minimum MSn values which means approx. zero which means they are not contributing to radiation. So this mode is responsible for generating a single band at 5.8 GHz frequency. Figure 7 shows the surface current and far field pattern of antenna-1, using this plot we can analyze the characteristics of mode-1, and mode-2. The radiation pattern and surface current of both modes are orthogonal at the same frequency.

For enhancing bandwidth, further circular patch antennas need to be modified in such a way bandwidth will increase, so that it converts into an elliptical shape as shown in Fig. 2b, using this concept the mode-1 and mode-2 separate some frequency difference as shown in Fig. 4. It also has the same λ_1, $\lambda_2 = 0$, and its MSn =1 at 5.75 and 5.88 GHz. Its corresponding surface current and radiation patterns are shown in Fig. 7.

After enhancing the bandwidth at 5.8 GHz by contributing mode-1 and mode-2, further ellipticals need to be modified to generate dual band characteristics. Figure 2c shows the elliptical patch along with a circular ring and both are connected by a strip line. In this structure mode-1, mode-2, and mode-3 are generated, because it has λ_1, $\lambda_2 = 0$ and its MSn = 1 at 3.5 GHz and 5.8 GHz respectively. Mode-1 contributes sub-6 band (3.5 GHz) and mode-2, mode-3 contribute WLAN band (5.8 GHz) as shown in Fig. 5.

Figure 8 shows the surface current distribution of antenna-3, which can analyze the surface current at mode-1, has a maximum at circular ring but a minimum at elliptical patch, according to current study we can say mode-1 only contributes at 3.5 GHz frequency, similar at 5.8 GHz frequency mode-2 and mode-3 has maximum current density and minimum current density at 3.5 GHz (Mode-1). Its corresponding radiation patterns of mode-1, mode-2, and mode-3 are shown in Fig. 8c, d. So according to surface current study and modification of conventional circular patch antenna, we can generate multi-band characteristics of antenna (Figs. 4 and 5).

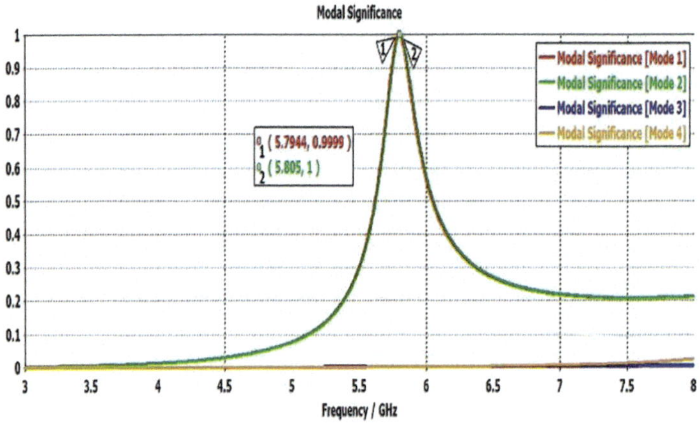

Fig. 3 Modal significant of antenna-1

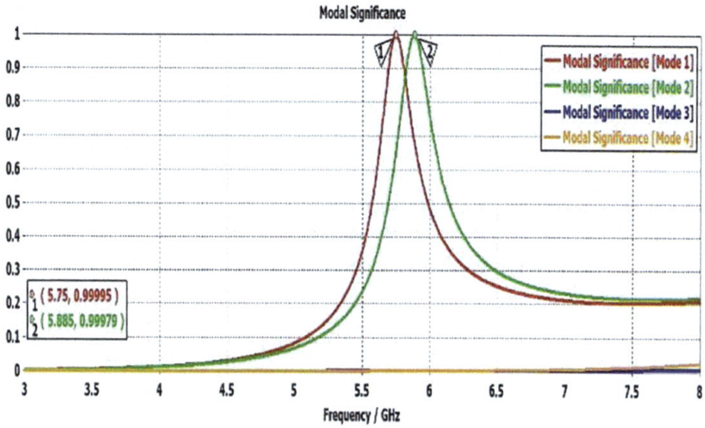

Fig. 4 Modal significant of antenna-2

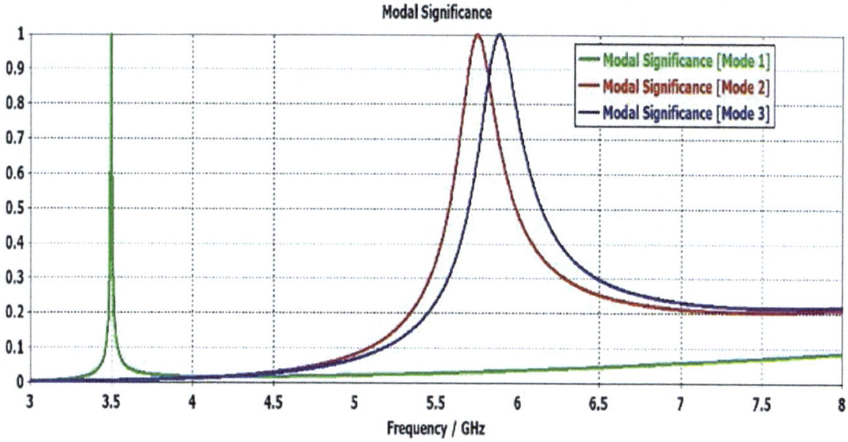

Fig. 5 Modal significant of antenna 3

2.2 3D FIT Analysis with Coaxial Feed Excitation

After analysis of characteristics mode analysis for multi-band antenna without any excitation. In this section the actual antenna analysis using coaxial feed excitation. The mode-1, mode-2, and mode-3 generated at 3.5 and 5.8 GHz frequency bands using CMA are excited using coaxial feed.

Figure 9 shows the step-by-step mode excitation using coaxial feed line. Initially mode-2, 3 simultaneously excited at 5.8 GHz as shown in Fig. 10 for antenna-1. After that, an elliptical shape antenna excited at the same frequency with enhanced bandwidth characteristics as shown in Fig. 10 antenna-2. The final proposed antenna

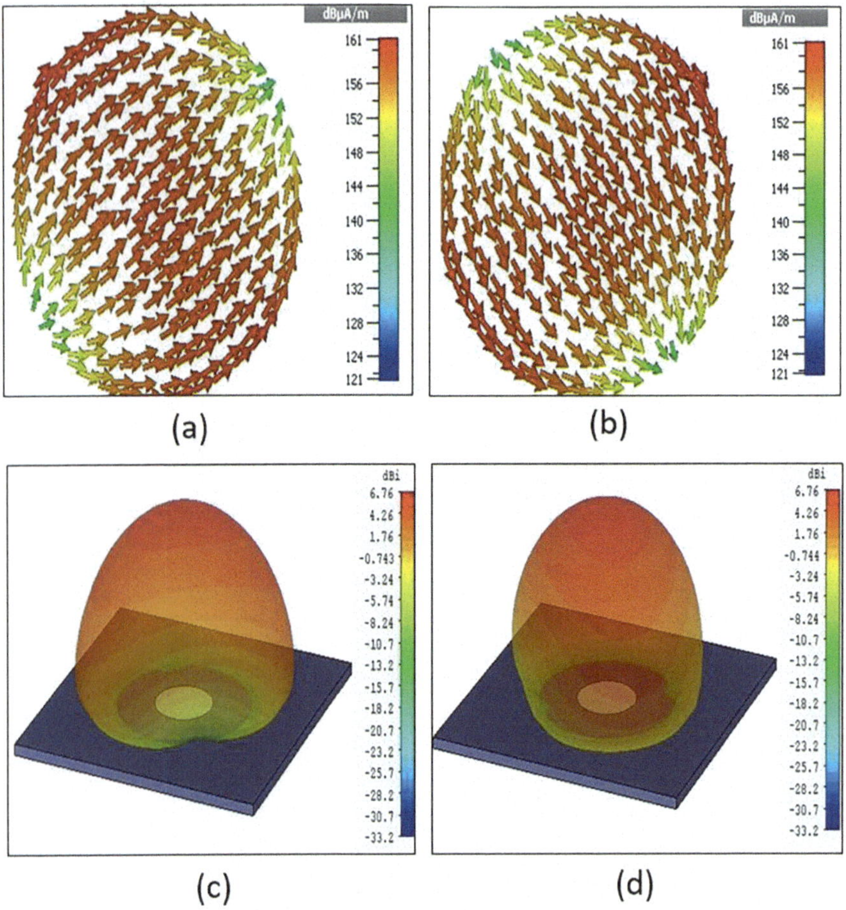

Fig. 6 Modal analysis of antenna-1, **a** surface current for mode-1; **b** surface current for mode-2; **c** far-field pattern of mode-1; **d** far-field pattern of mode-2

stricture excited mode-1, mode-2, and mode-3 simultaneously at 3.5 and 5.8 GHz frequency bands as shown in Fig. 10 antenna-3. The voltage standing wave ratio (VSWR) shown in Fig. 11, It has approx. unity values at 3.5 GHz and 5.8 GHz frequency band respectively and corresponding ration properties of antenna are shown in Fig. 12, it has 6.77 dbi gain at 3.5 GHz and 7.65 dBi gain at 5.8 GHz frequency band and hardware design is shown in Fig. 13 for the 3.5 GHz and 5.8 GHz frequency band.

The top view and bottom view of the fabricated dual band antenna and its simulation result are shown in Fig. 13 which resonates at 3.5 and 5.8 GHz.

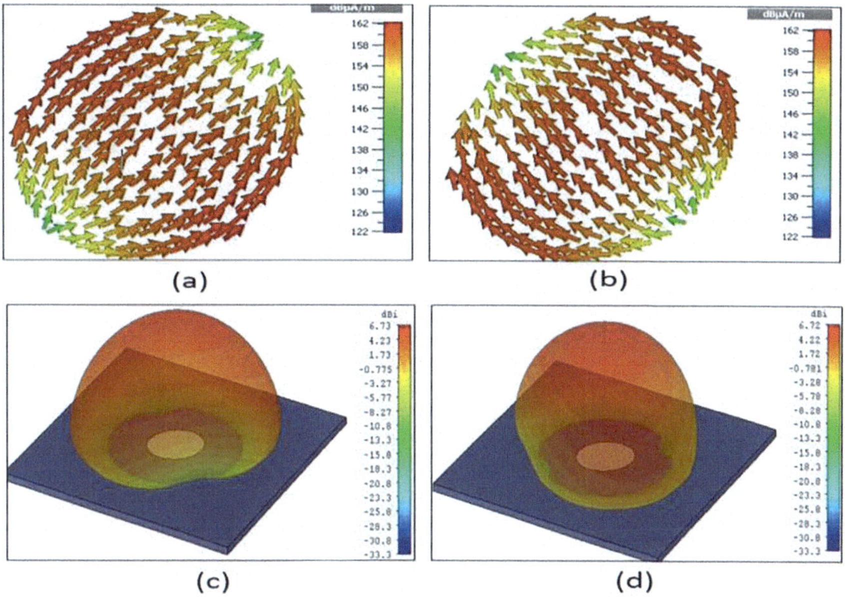

Fig. 7 Modal analysis of antenna-2, **a** surface current for mode-1; **b** surface current for mode-2; **c** far-field pattern of mode-1; **d** far-field pattern of mode-2

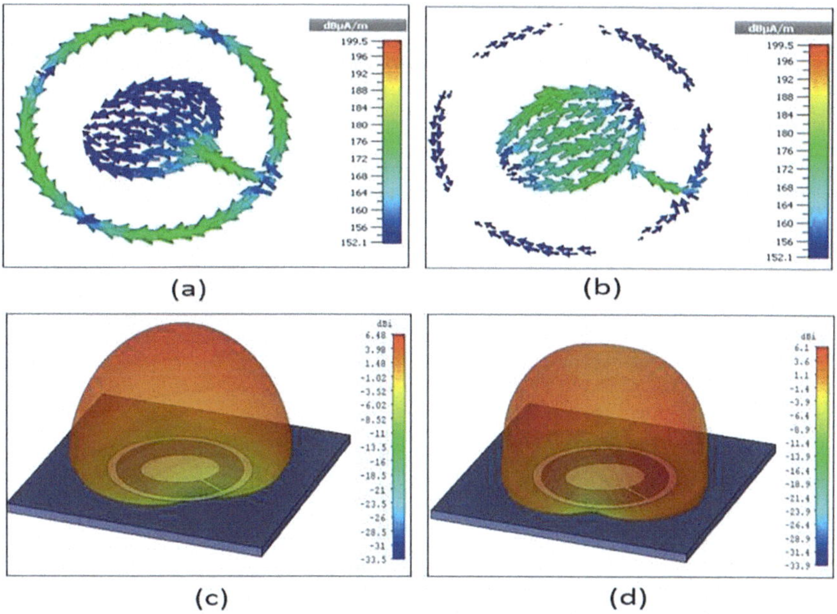

Fig. 8 Modal analysis of antenna-3, **a** surface current for mode-1; **b** surface current for mode-2; **c** far field pattern of mode-1; **d** far field pattern of mode-2

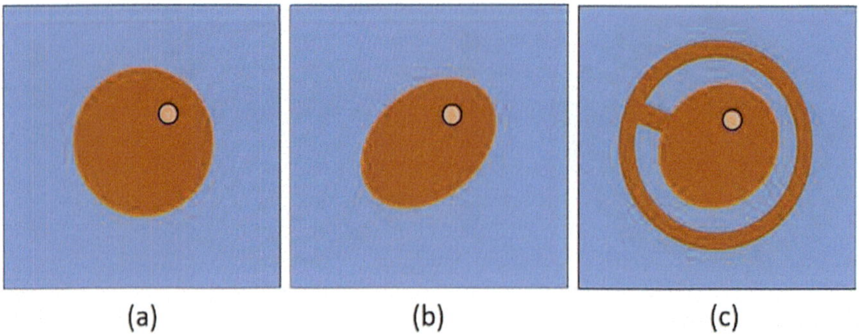

Fig. 9 FIT analysis of antenna, **a** antenna-1; **b** antenna-2; **c** antenna-3

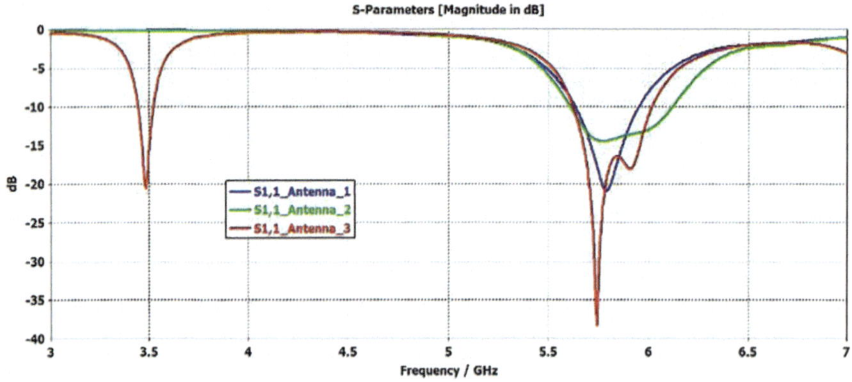

Fig. 10 S11 plot of antenna-1, antenna-2, and antenna-3

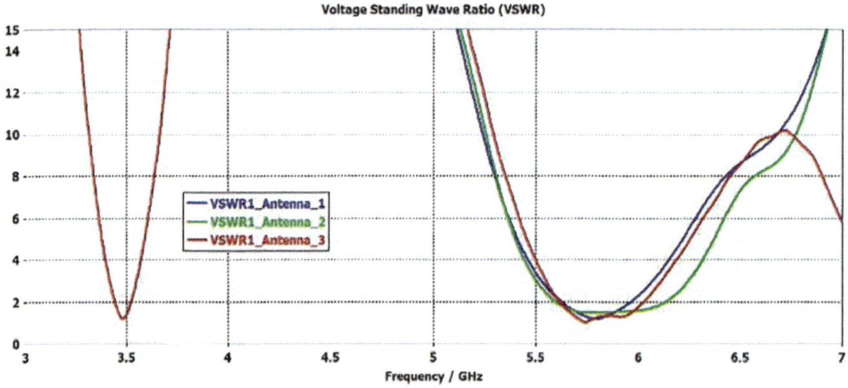

Fig. 11 VSWR plot of antenna-1, antenna-2, and antenna-3

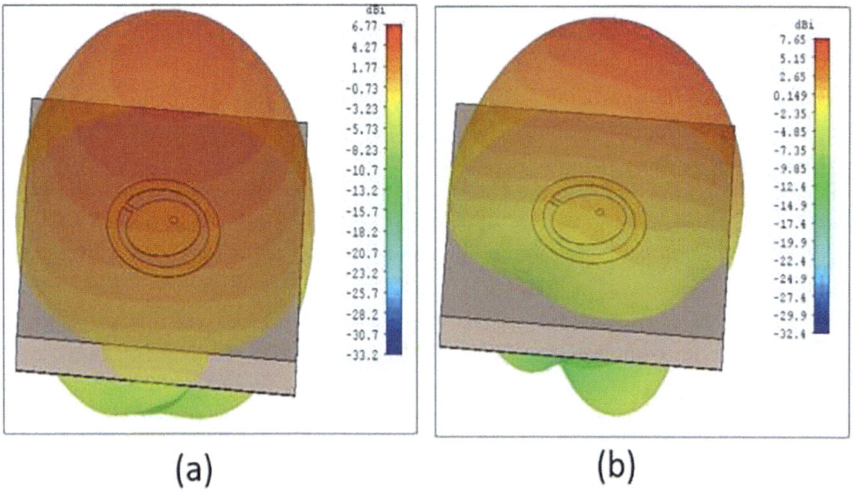

Fig. 12 Directive gain of propose antenna, **a** 3.5 GHz; **b** 5.8 GHz

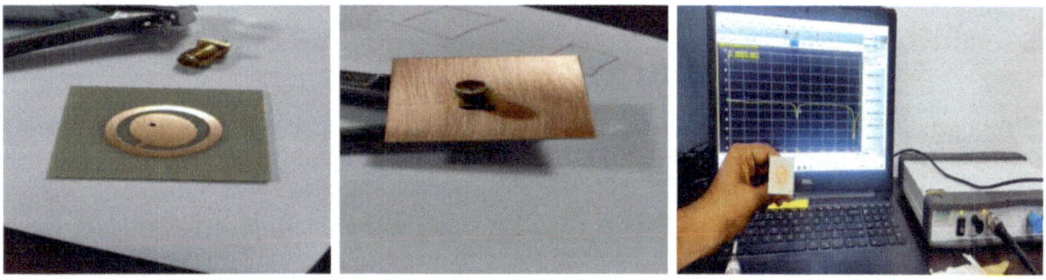

Fig. 13 Hardware design and testing of propose antenna at 3.5 and 5.8 GHz

3 Conclusion

This chapter demonstrated dual-band a novel antenna using characteristic modes analysis for 5G application resonating in frequency band Sub-6 and WLAN. The CMA theory provides a systematic design approach for the proposed antenna and a physical comprehension of electromagnetic concepts. The dual band antenna characteristics have been carried out using CMA and validated by full wave FIT simulation. The modes generated by antenna structure are excited by coaxial feed line with 50 Ω impedance which provides 6.77 dbi gain at 3.5 GHz and 7.65 dBi gain at 5.8 GHz frequency band. Proposed antenna has good broadside radiation and bandwidth properties. This antenna offers dual-band functionality and improved designing ease and reduces broader challenges within the field to minimize the antenna's overall size, simplify circuit integration, and optimize radiation quality thereby ensuring superior performance across the 5G sub-6 band and WLAN applications.

4 Future Perspective of Work

This antenna is developed using characteristic mode analysis (CMA) which is a strong tool for designing and conducting microwave circuits using a methodical approach like analysis, synthesis, and optimization is the characteristics mode theory. The integral surface equation and the MoM (Method of Moments) are used to quantitatively compute the characteristic modes of conducting structures which simplify circuit integration. This antenna operates at 3.5 GHz for the 5G sub-6 band and 5.8 GHz for WLAN. The achievement of dual-band characteristics is facilitated by incorporating an external circular ring within the elliptical structure. The external ring is specifically responsible for generating a lower mode at 3.5 GHz. A comprehensive full-wave analysis was conducted using the Finite Integration Technique (FIT) numerical approach, along with a coaxial feed, to thoroughly assess the performance of the proposed optimal design and ensure superior performance. The proposed antenna demonstrates desirable attributes for dual-band operation, making it well-suited for 5G and WLAN applications.

References

1. Yahia B, Ali M, Lila M (2022) Compact tri-band two-element MIMO slot antenna for 5G sub-6 GHz frequency bands. In: 2022 2nd international conference on advanced electrical engineering (ICAEE). IEEE, pp 1–6
2. Rani S, Tripathi K, Arora Y, Kumar A (2022) A machine learning approach to analyze cloud computing attacks. In: 2022 5th international conference on contemporary computing and informatics (IC3I), Uttar Pradesh, India, 2022, pp 22–26. https://doi.org/10.1109/IC3I56241.2022.10073468
3. Awan WA, Hussain N, Lee J, Hussain Q, Sufian MA, Kim N (2022) Stub loaded compact size tri-band antenna for ISM, 5G sub-6-GHz, IoT applications. In: 2022 Asia-Pacific microwave conference (APMC). IEEE, pp 539–541
4. Kumar L, Reddy BVR (2022) A compact and efficient triband antenna for Wi-Fi/WLAN/WiMAX/5G Mid-frequency band. In: 2022 international mobile and embedded technology conference (MECON). IEEE, pp 253–258
5. Sathishkumar N, Nandalal V, Natarajan R (2023) Design of a compact double-square-ring-shaped dual-band antenna for WiMAX/WLAN applications. Wirel Pers Commun 128(3):2073–2084
6. Alam MM, Azim R, Sobahi NM, Khan AI, Islam MT (2022) A dual-band CPW-fed miniature planar antenna for S-, C-, WiMAX, WLAN, UWB, and X-band applications. Sci Rep 12(1):7584
7. Singh J, Lohar FL, Sohi BS (2021) Design of circular polarized patch antenna for NaviC receiver applications. IOP Conf Ser: Mater Sci Eng 1033(1):012039. IOP Publishing
8. Ahmad I, Khan WUR, Dildar H, Ullah S, Ullah S, Mufti N, Hussien MI (2021) A pentaband compound reconfigurable antenna for 5G and multi-standard sub-6GHz wireless applications. Electronics 10(20):2526
9. Lohar FL, Sharma IB, Katewa V, Sharma MM (2021) T-shaped tri-band antenna based on characteristic mode analysis for satellite applications. Prog Electromagn Res C 115:65–81
10. Singh J, Lohar FL, Sohi BS (2021) Design of dual band millimeter wave antenna using SIW material for 5G cellular network applications. Mater Today: Proc 45:5405–5409

11. Suyan NK, Lohar FL, Solunke Y, Dhote C (2021) Design of compact size tri-band stacked patch antenna for GPS and IRNSS applications. In: Intelligent learning for computer vision: proceedings of congress on intelligent systems 2020. Springer Singapore, pp 501–509
12. Ullah S, Ahmad I, Raheem Y, Ullah S, Ahmad T, Habib U (2020) Hexagonal shaped CPW feed based frequency reconfigurable antenna for WLAN and sub-6 GHz 5G applications. In: 2020 International conference on emerging trends in smart technologies (ICETST). IEEE, pp 1–4
13. Dildar H, Althobiani F, Ahmad I, Khan WUR, Ullah S, Mufti N, Glowacz A (2020) Design and experimental analysis of multiband frequency reconfigurable antenna for 5G and sub-6 GHz wireless communication. Micromachines 12(1):32
14. Singh J, Lohar FL (2020) Frequency reconfigurable quad band patch antenna for radar and satellite applications using FR-4 material. Mater Today: Proc 28:2026–2030
15. Jain S, Rathee S, Kumar A, Sambasivam A, Boadh R, Choudhary T, Kumar P, Singh PK (2022) Prediction of temperature for various pressure levels using ANN and multiple linear regression techniques: a case study. Mater Today: Proc 56:194–199. https://doi.org/10.1016/j.matpr.2022.01.067
16. Kumar A, Parveen, Liu Y, Kumar R (eds) (2024) Handbook of intelligent and sustainable manufacturing: tools, principles, and strategies, 1st edn. CRC Press. https://doi.org/10.1201/9781003405870
17. Rani S, Kumar A, Bagchi A, Yadav S, Kumar S (2021) RPL based routing protocols for load balancing in IoT network. J Phys: Conf Ser 1950(1):012073. IOP Publishing
18. Kumar A, Rani S, Rathee S, Bhatia S (eds) (2023) Security and risk analysis for intelligent cloud computing: methods, applications, and preventions, 1st edn. CRC Press. https://doi.org/10.1201/9781003329947
19. Al-Tarifi MA, Faouri YS, Sharawi MS (2016) A printed 16 ports massive MIMO antenna system with directive port beams. In: 2016 IEEE 5th Asia-Pacific conference on antennas and propagation (APCAP). IEEE, pp 125–126
20. Yuan H, Wang C, Li Y, Liu N, Cui G (2016) The design of array antennas used for massive MIMO system in the fifth-generation mobile communication. In: 2016 11th international symposium on antennas, propagation and EM theory (ISAPE). IEEE, pp 75–78
21. Jain A, Yadav SK (2016) Design and analysis of compact 108 element multimode antenna array for massive MIMO base station. Prog Electromagn Res C 61:179–184
22. Al-Ajrawi Y, Rahhal J (2015) A simple antenna design for massive MIMO techniques. In: 2015 9th Jordanian international electrical and electronics engineering conference (JIEEEC). IEEE, pp 1–7
23. Hussain M, Mazher A, Chaudary E, Hussain B, Alibakhshikenari M, Falcone F, Limiti E (2021) Compact dual-band antenna with high gain and simple geometry for 5G cellular communication operating at 28 GHz and 44 GHz. In: 2021 XXXIVth general assembly and scientific symposium of the international union of radio science (URSI GASS). IEEE, pp 1–4
24. Ikram M, Nguyen-Trong N, Abbosh A (2019) Multiband MIMO microwave and millimeter antenna system employing dual-function tapered slot structure. IEEE Trans Antennas Propag 67(8):5705–5710
25. Li J, Zhang X, Wang Z, Chen X, Chen J, Li Y, Zhang A (2019) Dual-band eight-antenna array design for MIMO applications in 5G mobile terminals. IEEE Access 7:71636–71644
26. Moradi H, Eyvazi MR, Omidi A, Emadi PS, Pouradad E (2017) Design of the MIMO-UWB antenna for used in multi-frequency in wireless communication systems with the application of marine and aerial devices. In: 2017 IEEE 4th international conference on knowledge-based engineering and innovation (KBEI). IEEE, pp 0678–0682
27. Pouyanfar N, Ghobadi C, Nourinia J, Pedram K, Majidzadeh M (2018) A compact multi-band MIMO antenna with high isolation for C and X bands using defected ground structure. Radio Eng 27(3):686–693
28. Sundar R, Amir M, Subramanian R, Prabakar D, Giri J, Balachandran G, Ahmad F (2024) Spectral energy balancing system with massive MIMO based hybrid beam forming for wireless 6G communication using dual deep learning model. Heliyon e26085. https://doi.org/10.1016/j.heliyon.2024.e26085

Automation in Friction Stir Welding: A New Era of Fabrication Techniques

Meghnath Sen and Asit Baran Puri

Abstract The emergence of the fourth phase of industrial revolution has introduced significant advancements in manufacturing technologies and automated friction stir welding has become one of the cutting-edge technologies in this new era. It has emerged as a promising technique combining robotic precision with joint efficiency of friction stir welding to achieve superior welding quality and process flexibility. Robotic-friction stir welding offers several advantages, including enhanced joint quality, reduced production costs and time, and increased productivity. This article explores the technical aspects of robotic-friction stir welding, its applications across various industries, and the challenges encountered in its implementation. It highlights the potential of this technology to revolutionize manufacturing processes. Furthermore, this comprehensive article aims to provide a deeper understanding of the impact of robotic-friction stir welding process on modern manufacturing and its alignment with the principles of Industry 4.0 as well.

Keywords Robotic-friction stir welding · Industry 4.0 · Mechatronic system · Modern manufacturing

1 Introduction

With the advancement of time, the necessity for high quality products at low costs with specific and desired properties is increasing. This demand has led to a rapid industrial revolution that meets the high-performance and durability requirements of today's advanced applications. In response, new manufacturing processes have been developed that are faster and more efficient, compared to the conventional or traditional methods. Figure 1 illustrates the chronological progression of the industrial revolution. The first phase of industrial revolution or Industry 1.0 (I.R-1.0) began in the late eighteenth century, marking a pivotal shift to industrialized manufacturing

M. Sen (✉) · A. B. Puri
Department of Mechanical Engineering, National Institute of Technology Durgapur, Durgapur 713209, West Bengal, India
e-mail: meghnath001@gmail.com

A. Kumar et al. (eds.), *Advances in Mechatronics Systems*, Emerging Trends in Mechatronics, https://doi.org/10.1007/978-981-97-8472-1_5

[1]. In this phase, production shifted from manual labor to mechanized processes. The introduction of steam power to run machinery significantly increased the production capacity in the manufacturing sector. The second phase of industrial revolution or Industry 2.0 (I.R-2.0) started in the late nineteenth century with the introduction of electricity, assembly lines, and mass production techniques in manufacturing industries (e.g., steel and chemical industries) [2]. This revolution enhanced productivity, reduced production costs, and led to the rise of large-scale manufacturing enterprises. The assembly line significantly reduced the time and cost of manufacturing. In the mid-twentieth century, the third phase of industrial revolution or Industry 3.0 (I.R-3.0) introduced the use of computers and automation in manufacturing. In this era, there was a paradigm shift from analog to digital technology, increased automation of production processes, and the rise of the global supply chain. The use of PLCs (programmable logic controllers) and computer-aided design (CAD) systems revolutionized manufacturing processes [3]. The implementation of mechatronics and robotics made the manufacturing process more precise and accurate, decreased production time and unit cost. The fourth phase of industrial revolution or Industry 4.0 (I.R-4.0), which began in the twenty-first century, introduced the cyber-physical system into the manufacturing process. This industrial revolution enhanced efficiency, flexibility, customization in manufacturing through the use of real-time data. It utilised advanced analytics effectively for predictive maintenance and optimization of production processes. I.R-4.0 represents a shift towards more interconnected, automated, and intelligent manufacturing environments [4–10]. The key innovations are Internet of Things (IoT), big data and analytics, machine learning, artificial intelligence (AI), advanced robotics, autonomous systems, additive manufacturing, cloud computing, and so on, as shown in Fig. 2. Each industrial revolution has built upon the innovations of the previous one, leading to increasingly sophisticated and efficient production methods. I.R-4.0, in particular, represents a merging of digital and physical technologies, fostering an era of unprecedented connectivity and intelligence in manufacturing.

Mechatronics, being an interdisciplinary field, primarily focuses on integrating computer science, mechanical, electrical, and control engineering to design and create innovative and efficient systems, as illustrated in Fig. 3. It allows for the development of intelligent machines and systems that can perform complex tasks with precision and accuracy [11]. The core components of mechatronics include sensors, actuators, microcontrollers, software, algorithms, and so on [12]. These elements work together to enable machines to work precisely, collect and analyze real-time data, make decisions, and execute actions. By combining the principles and techniques from multiple engineering disciplines, mechatronics enhances the functionality, performance, and versatility of systems, driving innovation and efficiency in modern technology [13, 14]. Mechatronics is applied in different important sectors, including manufacturing, robotics, automotive, aerospace, and healthcare, leading to advancements such as automated production lines, intelligent robots, advanced vehicle systems, medical devices, and so on [15–22].

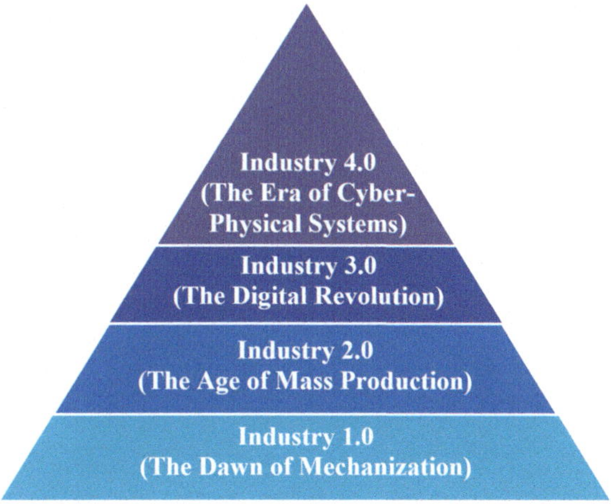

Fig. 1 Chronological progression of the industrial revolution

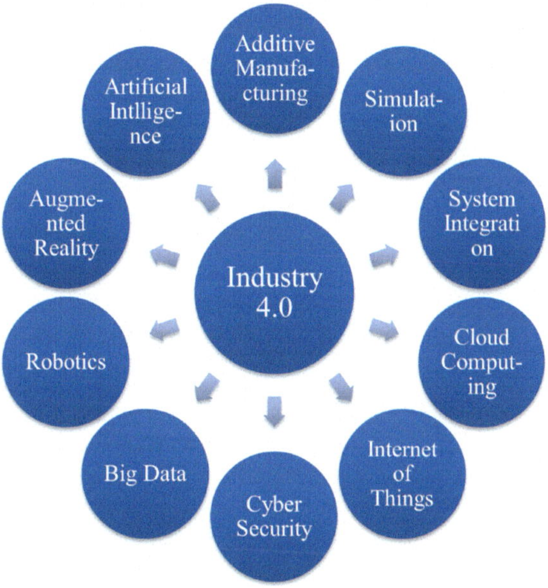

Fig. 2 Core components of Industry 4.0 (I.R-4.0)

The convergence of mechatronics and I.R-4.0 represents a transformation in the modern manufacturing process. It enables the development of highly automated intelligent machines and production systems that can manufacture even complex products accurately with sound precision within very little time. Mechatronics is at the heart of modern robotics and automation technologies, where mechanical components are controlled by software and networked together, allowing for enhanced monitoring, control, and optimization. Robot-assisted manufacturing processes are useful in terms of human safety, minimizing waste, reducing energy consumption, and

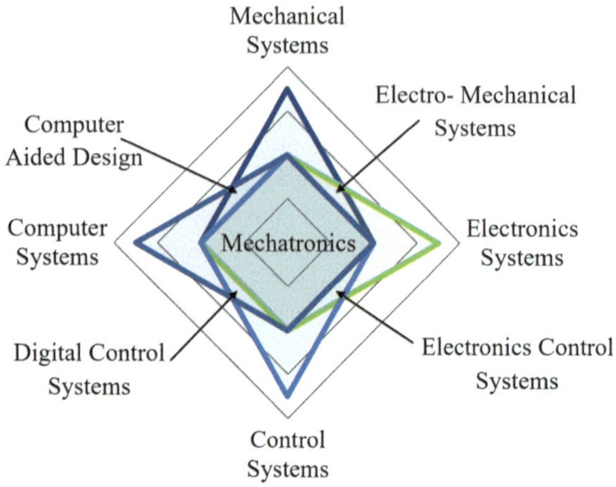

Fig. 3 Elements of mechatronics systems

optimizing resource utilization. The robotic-friction stir welding (R-FSW) makes the I.R-4.0 more realistic, which contributes significantly to the evolution of smart manufacturing. R-FSW is one of the significant manufacturing processes developed within I.R-4.0, which is currently a fast-growing process of production.

This article emphasizes the R-FSW process, detailing its benefits, applications, and challenges. By exploring the advancements and impacts of R-FSW, this book chapter highlights how this technology enhances manufacturing efficiency and quality supporting the broader goals of I.R-4.0.

2 Background of Friction Stir Welding (FSW) Process

Welding is a fabrication process utilized to join similar or unlike materials (typically metals, thermoplastics, etc.) together. The process involves melting the parent materials (work materials) using a filler material optionally to form a weld pool. As the weld pool cools and solidifies, it forms a strong bond between the materials. The process involves the application of heat to the materials being joined, causing them to melt and fuse together. It is essential to the industries where the integrity and reliability of joints are important. Welding serves various purposes, including the fabrication of structures, components, and assemblies, as well as repair and maintenance activities. It is a versatile and crucial process that enables the creation of complex and durable structures, contributing significantly to the advancement of modern technology and infrastructure. Traditional welding methods or techniques, i.e. arc welding, oxy-acetylene gas welding, resistance welding etc. have been the backbone/pillars of the manufacturing industries, each presenting distinctive advantages and uses. However, these methods typically involve the melting of base materials, which can result in thermal gradients, residual stresses, and potential metallurgical changes within the welded joints. The choice of welding technique (process) depends on various key factors, including the materials being joined, the desired strength and quality of the weld required, and the particular applications [23–32].

Amidst the numerous advantages of conventional welding methods, challenges arise while welding materials with low melting points. Low melting point materials tend to flow more readily under high heat flame, which can result in excessive material displacement, overheating, and inadequate mixing during welding. This phenomenon can lead to incomplete joint formation, defect formation, uneven distribution of material, and compromise the quality or standard of the weld. In addition, controlling temperature during welding is challenging in traditional welding methods. High-strength materials, i.e. aluminium alloys are challenging to join or weld using traditional fusion welding methods or techniques due to their low melting points. Such materials are called "hard-to-weld" materials [33–35].

To overcome such issues, a solid-state joining, called 'friction stir welding' (FSW), was first developed by The Welding Institute (TWI) in the year 1991, which revolutionized the field of joining materials. It has emerged as an innovative alternative joining process, offering distinct advantages and opening new frontiers in materials joining. Solid-state welding methods involve the joining of work materials without reaching their melting points, thereby avoiding the issues associated with fusion-based welding processes. Although it was developed for aluminium alloys initially, FSW has been adapted for various other materials, including magnesium, copper, steel, titanium, and even some non-metallic materials. It is beneficial for numerous industrial uses and applications, ranging from aerospace and automotive industries to shipbuilding and construction works, and many more [36–38].

2.1 Principle of Operation

In FSW process, a non-consumable tool rotates that helps in heat generation and mechanically stirs the parent materials together, resulting in sound weld joints with minimal distortion and superior mechanical properties. The tool has two main parts: a pin or probe with a smaller diameter and a shoulder with a larger diameter. During welding, the tool is inserted into the weld line between the clamped workpieces. The pin completely penetrates into the work materials, while the tool shoulder touches the work material's surface. The friction between the work materials and tool produces heat and softens the parent materials without reaching their melting point. This enables plastic deformation and subsequent bond formation. The fundamental concept and working of the FSW process are detailed below [35] along with an illustration in Fig. 4.

- **Heat Generation (Plunging Stage)**: Initially, the tool is gradually fed into the weld line between the tightly clamped parent materials under a constant axial load. This continues until the shoulder of the tool comes in contact with the surface of the workpieces. Frictional heat is then generated between the tool and the work-material due to the relative frictional motion and pressure applied. The process

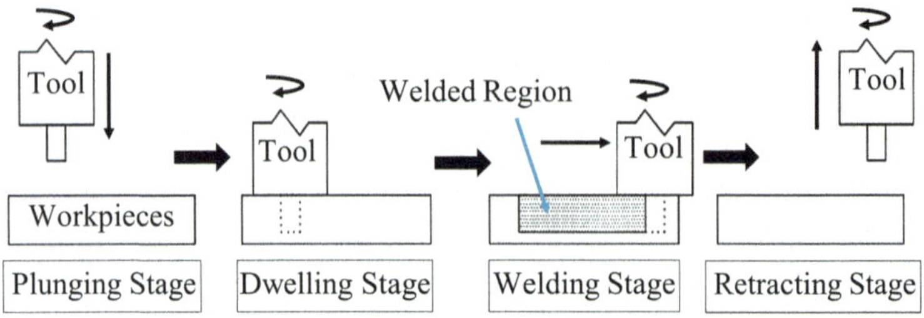

Fig. 4 Steps of FSW process

proceeds further until the temperature at the contact interface rises sufficiently to soften and plasticize the work materials.

- **Plasticization (Dwelling Stage)**: The generated heat softens the parent materials and converts the parent materials into semi-solid state without reaching their melting point. This softening causes the material to become plastically deformable, allowing it to be stirred and mixed together. During the dwell period, the tool rotates without translator motion, allowing the temperature to steadily rise until it reaches the desired level.
- **Material Movement and Joint Formation (Welding Stage)**: As the tool spins and travels along the line of weld, it forces the semi-solid softened material to flow. This movement effectively stirs the material together, creating a solid-state bond.
- **Cooling and Solidification (Retracting Stage)**: After the tool passes through, the welded material cools and solidifies, forming a strong weld joint. Finally, the tool is withdrawn from the work material.

2.2 *Friction Stir Welding Machine*

In the field of FSW, three main types of machines are recognized as suitable options for performing the process, each offering distinct advantages and suitable applications. These are:

- Conventional machine tools (milling machines) [34, 39, 40]
- Dedicated or custom-built FSW machines [41]
- Robot-assisted FSW machine [42–46]

2.2.1 Conventional Machine Tools

Conventional machine tools (i.e. milling machines, lathe machines) have been adapted primarily to perform FSW processes due to their availability and versatility. These machines are prevalent in manufacturing industries and used for different machining purposes. However, such machine tools are not initially designed for

FSW, but they can be modified to carry out the process effectively, providing a cost-efficient and accessible option for many manufacturers. Vertical milling machines are ubiquitous among them, where vertical spindle can be equipped with FSW tools very easily. The main advantages of using conventional machine tools for FSW are:

- These machines are multipurpose and can switch between traditional machining tasks and FSW, maximizing their utilization.
- These machines are capable of handling various workpiece sizes and materials.
- Machine tools are widely available in many manufacturing units, and do need any special type of training for operation.
- Basic modifications, such as adding a custom FSW tool and fixture (adequate clamping system) for workpiece holding, can convert a milling machine into an FSW-capable machine. This can be a cost-effective solution also.

However, challenges associated with such machine tools are:

- Lack of advanced control systems, which are necessary for optimal parameter control (e.g., spinning velocity of the tool, traverse speed, and axial load or force).
- Upgrading the control systems can be complex and costly.
- Lack of precision of machine tools as conventional transmission system (belt-pully, rack-pinion, etc.) is used.
- Limitations of welding length, as machine table size is limited.
- Required axial thrust is not achieved during the process.

Conventional machine tools provide an economical solution for performing FSW, making them ideal for small to medium-sized enterprises. They are capable of joining both thick and thin workpieces of any length effectively.

2.2.2 Dedicated or Custom-Built FSW Machines

Dedicated or custom-built FSW machines are specifically designed and optimized to perform the FSW process. These machines provide superior control, precision, and efficiency, making them ideal for applications that require high-quality welds and consistent production outcomes. These machines are used to join lightweight, high-strength components such as aircraft fuselage panels, car bodies, fuel tanks, etc., reducing the overall weight of the vehicles and improving fuel efficiency. It ensures high-integrity welds that are crucial for safety and performance. They are ideal for producing complex components and assemblies with high precision. Such machines have advanced controlling systems, integrated cooling systems, and force and temperature measurement systems. The main advantages of such custom-built machines are:

- These machines ensure consistent and high-quality welds with minimal defects.
- Required axial thrust can be achieved during the process.
- It has precise control over key welding inputs such as the spinning velocity of the tool, traverse speed, tilt angle, axial load or force, etc.

- It has integrated systems for real-time monitoring and adjustment of process parameters, ensuring the welding conditions and quality remain optimal.
- These machines have sound rigidity to maintain precise tool positioning and minimize deflection during welding.
- They have advanced programming capabilities to handle complex welding paths and multi-axis movements.
- These machines reduce setup times and faster welding cycles compared to conventional machine tools.
- Advanced automation reduces the need for manual intervention, lowering labor costs and increasing repeatability.

However, these machines have some limitations, including higher cost, trained operators, larger space, and higher maintenance costs.

2.2.3 Robot-Assisted FSW Machine

Robot-assisted FSW machines offer a versatile, efficient, and highly precise solution for performing the FSW process. Their ability to automate the process, adapt to complex geometries, and maintain consistent weld quality makes them an invaluable asset in modern manufacturing. Despite the initial investment and technical challenges, the benefits of enhanced productivity, improved safety, and long-term cost savings make robotic-FSW an attractive option for many industries. These machines are ideal for welding lightweight, high-strength alloys or materials used in aerospace, automobile, marine, and different manufacturing sectors. The key advantages of robot-assisted FSW machines are:

- Robots equipped with multiple axes of motion can navigate complex welding paths and orientations, making them ideal for linear as well as non-linear joints.
- Industrial robots can be integrated into automated production lines, increasing efficiency and throughput while reducing manual labor requirements.
- Integration with I.R-4.0 technologies, i.e. IoT and AI, enables real-time monitoring, adaptive control, and data-driven optimization of the welding process.
- Robots can operate in smaller, more confined spaces compared to larger dedicated machines.
- Robots can operate continuously without fatigue, significantly increasing production rates.
- Robots can perform continuous welding without any size limits.

However, there are some limitations, including the higher initial investment in robotic systems, which covers the cost of robots, essential software, and necessary training. Programming and maintaining industrial robots require specialized skills and knowledge. Additionally, these systems often demand higher levels of maintenance.

2.3 Key Components of FSW Machines

- **FSW Tool**: The heart of the process is usually made from a hard and wear-resistant material like tool steel or tungsten carbide. The tool's geometry is critical for efficient heat generation and material flow.
- **Clamping System**: Ensures that the workpieces remain securely in place during welding, preventing movement that could result in a defective weld.
- **Spindle and Motor**: Provides the tool with the necessary rotational speed and torque.
- **Control System**: Manages the process inputs such as the spinning velocity of the tool, travel speed, tilt angle, and axial load or force to ensure optimal welding conditions.

2.4 Advantages

The FSW process reveals several notable advantages and benefits compared to conventional or traditional welding methods. Some of the key advantages are:

- **Higher Joint Strength**: FSW produces welds with sound mechanical properties, including higher strength, due to the absence of defects concerning the fusion phenomenon.
- **Low Distortion**: Because FSW is a solid-state joining method, it generates minimal distortion and residual stresses in the workpieces, making it suitable for joining thin and lightweight materials.
- **Environmentally Friendly**: FSW produces no fumes, gases, or arc emissions, making it environmentally friendly and safe for operators.
- **Wide Material Compatibility**: FSW can be used to join a broad range of different materials, including aluminium, magnesium, copper, steel, titanium, composites, and so on.

2.5 Applications

FSW is employed across various industries due to its reliability and efficiency [47–50]:

- **Aerospace**: Used for manufacturing lightweight, high-strength components such as fuselage panels and fuel tanks.
- **Automotive**: Ideal for joining aluminium components in car bodies, improving fuel efficiency by reducing weight.
- **Shipbuilding**: Ensures strong, durable joints for aluminium hulls and superstructures.

- **Railway**: Used in the construction of high-speed train carriages and other structural components.
- **Electronics**: Joins heat sinks, enclosures, and other components where thermal management is critical.

3 Robotic-Friction Stir Welding

3.1 Basic Principle

Robotic-friction Stir Welding (R-FSW) is a superior welding or joining technique that merges the friction stir welding process with the precision and flexibility of robotic automation, as illustrated in Fig. 5. The key components of the process are sensors, actuators, end effectors, and controllers [44–46, 51–55]. In the R-FSW process, the base provides stability and support for the robotic arm. Links/arms are connected through joints (shoulder, elbow, and wrist) to ensure the stability of the system/structure. Each joint offers a degree of freedom, allowing the arm to bend, twist, and rotate. Actuators are devices that convert energy into motion, driving the movement of the joints. Common types of actuators are electric motors, hydraulic cylinders, and pneumatic actuators. The end effector, attached to the robotic arm's end, is designed to grip and hold the FSW tool precisely. The controller serves as the central intelligence of the robotic arm, processing input from sensors and sending commands to actuators. It runs software that directs the arm's movements and operations. Sensors are used to provide feedback to the robotic arm, enabling it to interact intelligently. Force, motion, position, and temperature detection sensors are the most common types of sensors used in this process. Figure 6 represents the complete cycle of the R-FSW process. The controller unit is the central hub, managing the entire operation, controlling welding parameters, and monitoring real-time data to ensure precision. The robot arm, equipped with joints and actuators, accurately moves and positions the FSW tool. This tool, comprising a tool pin and shoulder, generates the frictional heat necessary for welding.

Fig. 5 Schematic diagram of R-FSW process

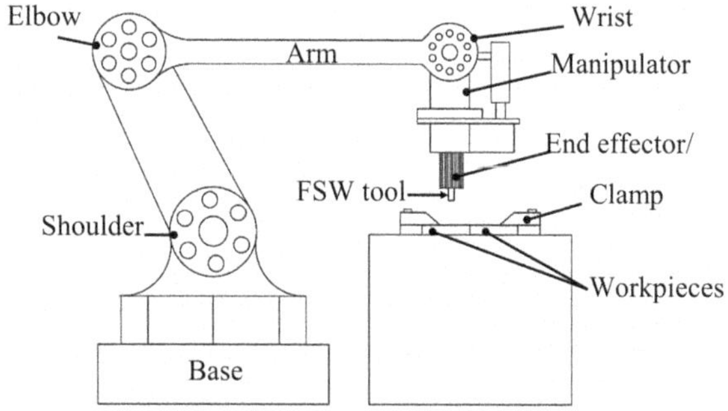

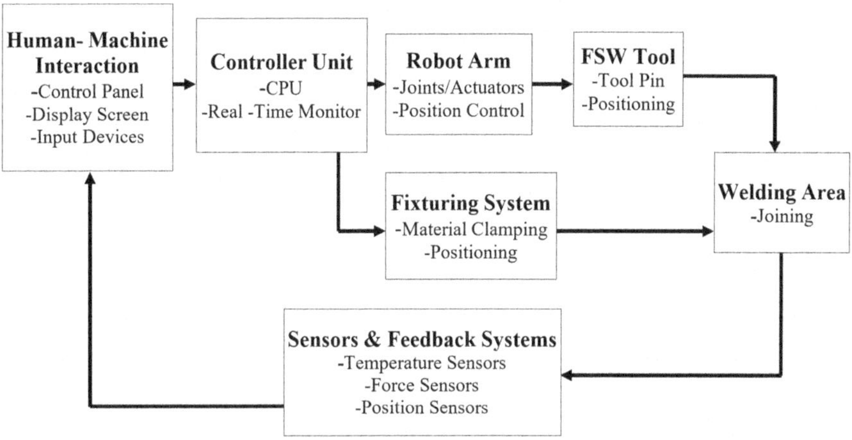

Fig. 6 Block diagram of R-FSW process

The fixture system securely holds the workpieces in place, ensuring they are accurately aligned and positioned for welding. The actual welding takes place in the welding area, where the workpiece material and joint line are located. Different sensors, including a feedback system, continuously monitor parameters such as temperature, force, and position, providing real-time data to the controller for process adjustments. Operators interact with the system through the human–machine interface, which includes a control panel, display screen, and input devices. These components collectively enable a precise, efficient, and safe R-FSW process.

3.2 Types of Robots Used in R-FSW

A notable advancement in the realm of FSW is the introduction of robotic systems, specifically industrial robots. In recent years, technological developments have enabled the use of robotic systems with advanced automation, significantly enhancing productivity in a shorter time duration. Most of these robots possess five or six degrees of freedom (DOFs), which is essential for performing multiple tasks with high precision. These advancements in R-FSW not only improve efficiency but also open up new possibilities for complex and precise welding applications across various industries. Commonly, two types of robots are used in R-FSW systems, each with unique features based on their application, as illustrated in Fig. 7:

(i) Articulated robotic arms [42–46, 51–53, 56–59]
(ii) Parallel and hybrid robots [51, 60–65]

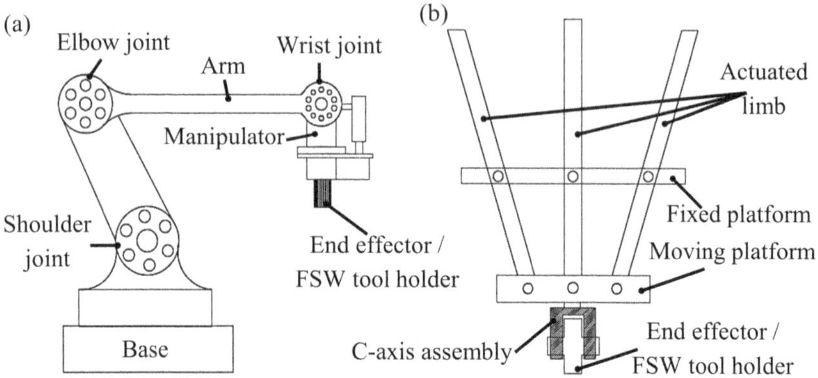

Fig. 7 **a** Articulated robotic arm, **b** parallel and hybrid robot

3.2.1 Articulated Robotic Arms

Articulated arm robots feature a series of rotary joints, providing a high degree of flexibility, as shown in Fig. 7a. Their design allows for a wide range of motion, making them suitable for welding complex and irregularly shaped workpieces. These robots are ideal for applications, where precision is crucial, enabling multi-directional welding paths and handling various positions and orientations, which enhances their versatility in industrial applications.

Several studies have demonstrated the effectiveness of articulated arm robots in R-FSW processes. Patel et al. [42] developed a lightweight assembly of aluminium alloys, used in battery frames of electric vehicle industries using R-FSW process. To join aluminium plates (thickness 3 mm each) in lap configuration, a 5-axis robot (ESAB Rosio), equipped with a force sensor and temperature measurement system was employed. This setup produced defect-free welds with good intermixing of materials and enhanced mechanical properties. Zhao et al. [43] found the ZK-500 industrial robot (DOFs-5) advantageous for the R-FSW process due to its multiple DOFs and flexibility. In exciting research by Soron et al. [44], a modified ABB IRB 7600-500 robot, equipped with a force sensor and force controller was used to join 6063 aluminium plates of 3 mm thickness, exhibiting excellent stiffness and good payload capabilities. Mendes et al. [45, 46] successfully used motorman ES165N robotic arm to join 6 mm thick polymer (ABS) plates, concluding that the use of robotic arm is more flexible, cost-saving, and time-efficient. Bres et al. [52] employed a KUKU KR500 MT robot to successfully join 2.3 mm thick aluminium 6013-T4 alloys. Further, the ABB 7600-500 (ESAB Rosio) 5-axis industrial robot was used to join volvo car's engine cylindrical blocks (made of aluminium), demonstrating the ability to weld highly complex geometries with lower cost and less time [53]. Wanjara et al. [56] joined 3.18 mm thick aluminium 6061-T6 sheets using a robot (KUKA KR500) for R-FSW. Guillo et al. [57] welded 3 mm thick aluminium 5754-H22 sheets using FANUC S900iB/400RJ3iB robots successfully. Wang et al. [58]

developed a dynamic model for axial force simulation in R-FSW process using a KUKU KR500-2MT industrial robot (DOFs-6).

The major advantages of articulated robots include high flexibility, high repeatability, good decision-making capability, and lower cost compared to other robots. However, these robots can only weld a limited range of materials and thicknesses, due to their moderate load capacity and low stiffness. At a high load, their accuracy decreases, although they can weld thin materials with higher precision.

3.2.2 Parallel and Hybrid Robots

Parallel and hybrid robots, also known as 'Tricept/TriMule robots', utilize a different configuration with multiple linear actuators working in parallel to control the orientation and position of the end effector, as presented in Fig. 7b. This design provides high stiffness and precision, making these robots ideal for jobs/tasks requiring a high degree of accuracy and stability. They are particularly effective for applications requiring consistent, high-quality welds on components with moderate thickness.

Xiao et al. [60] conducted an R-FSW experiment to develop S-shaped joint using a TriMule 800 robot (5-axis) with controlling constant plunge depth, achieving successful welding. Additionally, Yue et al. [61] performed an R-FSW experiment to join 5.5 mm thick aluminium 2024-T4 sheets using a TriMule robot (5-axis) successfully, predicting the stiffness of the robot. Zaho et al. [62] developed a model for controlling nonlinear axial force using an ABB tricept robot IRB 940 (DOFs-6), equipped with an architecture control system and force sensor. Further, Li et al. [65] confirmed the accuracy of a multi-parameter sensing model using a TriMule 800 (5-axis) robot, achieving high precision for process inputs, i.e. tool tilt angle, plunge depth, and others respectively.

3.3 Systems in R-FSW

3.3.1 Force and Torque Control

Force and torque control in R-FSW is a crucial aspect that ensures the quality and consistency of the welds produced [66–74]. By sensing forces and torque, the system gains monitoring capabilities that provide insights into weld properties and expand control capabilities, making R-FSW more adaptable to industrial robots. Additionally, force sensing enables the implementation of joint tracking techniques.

One of the primary objectives of force control in R-FSW is to maintain a constant and appropriate force exerted by the welding tool onto the work-material surface. This force facilitates the generation of frictional heat and plasticization of the material, which is essential for achieving a strong metallurgical bond without melting. By

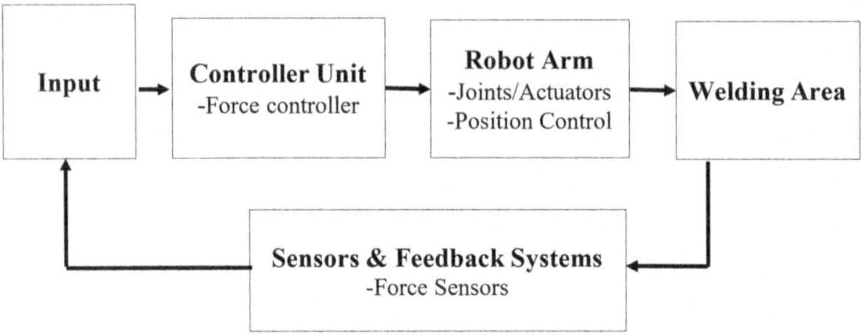

Fig. 8 Block diagram of force feedback system

controlling the applied force, the robot can ensure uniform material flow and intermixing at the weld interface, resulting in improved joint integrity and mechanical properties.

In R-FSW the primary forces are the axial force, the traversing force, and the torque. These forces directly influence the quality and integrity of the weld, making their precise control essential for achieving desired outcomes. However, there are drawbacks associated with FSW such as the requirement of adequate force during the process. The higher force can lead to challenges such as unpredictable errors in tool positioning, while lower force can develop poor weld joints with defects.

The torque equation, as described in literature [51] are:

$$\text{Torque (T)} = \text{Torque on (tool shoulder} + \text{side face of tool pin} + \text{bottom face of tool pin)}$$

$$T = \sigma\left(\int_r^R 2\pi r dr + 2\pi r^2 l + \int_0^r 2\pi r^2 dr\right) \tag{1}$$

where 'r' is the radius of the tool pin, 'l' is the length of the tool pin, 'R' is the radius of the tool shoulder, and 'σ' is the shear flow stress.

Force control in R-FSW is typically achieved through closed-loop feedback system, as illustrated in Fig. 8. Sensors, such as force transducers are integrated into the robotic system to continuously measure the force exerted by the welding tool. This real-time feedback is used to adjust the robotic motion and maintain the desired force level throughout the welding process.

3.3.2 Plunge Depth Control

Plunge depth control is an important and crucial aspect of R-FSW that involves regulating the penetration depth of the welding tool into the work-material. It refers to the ability to precisely control how deep the tool plunges into the material during the welding process [60, 75]. Maintaining consistent plunge depth is essential for achieving uniform weld quality and mechanical properties. It ensures that the tool effectively mixes and consolidates the material along the entire depth of the weld seam, leading to a strong and defect-free joint.

Several factors influence plunge depth control, including the spinning velocity of the tool, the traverse speed of the robot, and the material properties. R-FSW systems may employ closed-loop feedback control mechanisms, such as depth sensing, to continuously monitor and adjust the plunge depth in real-time. Effective plunge depth control enhances the repeatability and reliability of the welding process, resulting in high-quality welds with consistent properties.

3.3.3 Motion Control

Motion control is a fundamental aspect of R-FSW that governs the movement of the welding tool and the robotic arm during the welding process [76, 77]. It encircles the precise coordination of various motion parameters, such as the spinning velocity of the tool. etc., to achieve the geometry and quality of the desired weld. In R-FSW, motion control plays an important and critical role in determining the welding speed along the spinning velocity of the tool, and plunge depth into the workpiece material. These parameters must be carefully and precisely controlled to ensure proper material plasticization, optimal heat generation, and effective mixing of the materials without causing defects and excessive distortion. Several factors influence motion control in R-FSW, including the material properties, material thickness, joint configuration, and desired weld characteristics. Mechatronics systems may employ sophisticated control algorithms and feedback mechanisms to adjust motion parameters dynamically based on real-time sensing data. Effective motion control enables the robot to accurately follow simple as well as complex welding paths, adapt to changes in material properties and joint geometry, and maintain consistent weld quality across different workpieces. It also contributes to improved productivity, as precise motion control allows for faster welding speeds while maintaining weld integrity. Overall, motion control is essential for achieving high-quality and reliable welds in R-FSW applications. By ensuring precise and coordinated movement of the welding tool, motion control systems contribute to the efficiency, repeatability, and performance of the welding process across various industrial sectors.

3.3.4 Stiffness

Stiffness in R-FSW refers to the ability of the robotic system and the welding tool to resist deformation and maintain stability during the welding process. It plays an essential role in ensuring the accuracy, precision, and quality of the welds produced. When a force is exerted on the robot's end-effector, both joints and links undergo elastic deformation. This can lead to misalignment of the total system. In R-FSW, stiffness is essential for maintaining the proper alignment and contact between the work materials and the welding tool. A rigid and stable end effector prevents excessive vibrations and deflections, which can lead to weld defects. Stiffness is particularly important in controlling the applied forces and torques during the welding process. A rigid robotic system can accurately regulate the forces exerted by the welding tool, ensuring consistent penetration depth and material flow. It also helps control the torque applied to the tool, maintain stable rotational motion, and prevent tool chatter. Moreover, stiffness influences the dynamic response of the robotic system to changes in process conditions and external disturbances. A highly stiff system responds quickly and accurately to these variations, resulting in improved process stability and reliability. To achieve optimal stiffness in R-FSW systems, various factors must be considered, including the design and construction of the robotic arm, end effectors, the rigidity of the tooling and fixturing, and the material properties of the workpiece. Additionally, real-time feedback mechanisms can be employed to enhance stiffness control during the welding process.

3.4 Advantages

The advantages of R-FSW processes include:

- Robotic FSW systems can achieve precise control over welding inputs, i.e. spinning velocity of the tool, traverse speed, and plunge depth, resulting in consistent and high-quality welds.
- Robots can be programmed to weld complex shapes and contours, making them well-suited for a broad industrial application across various sectors.
- The utilization of automation in R-FSW process leads to enhanced productivity and reduced labour costs and manufacturing time compared to manual welding methods.
- Robotic FSW systems can maintain consistent weld quality and integrity, reducing the chances of defects.
- By automating the welding process, robotic FSW can improve workplace safety by reducing the need for manual labor in the workplace.
- Robotic FSW systems can be easily reprogrammed to accommodate changes in production requirements through welding tasks.
- With precise control over welding parameters, R-FSW processes can produce welds with uniform properties and dimensions, ensuring repeatability and reliability.

Overall, robotic FSW offers numerous advantages and benefits over traditional welding methods, making it a preferred choice for many industrial applications.

3.5 Limitations

Although, there are so many advantages of R-FSW process, there are some limitations also. The major limitations are:

- Robotic arms exhibit poor stiffness during welding of thick sheets and hard materials, where large axial force is required.
- Initial installation cost is high for R-FSW system.
- Regular maintenance is required to avoid systematic errors, leads to more operational cost.
- Position error is very common in R-FSW due to larger axial force applied in tool.

3.6 Applications

The applications of R-FSW are diverse and span across numerous important industries [42, 51–53, 61, 63, 78–96]:

Automotive industries: The main objective of automotive industries is to produce lightweight and durable components, that improve fuel efficiency and reduce emissions. The industry is undergoing a significant transformation towards green transition and sustainable development, particularly with the increasing adoption of battery electric vehicles. A crucial aspect of this change is the requirement to reduce vehicle weight, enhance energy efficiency, and minimize emissions. Key components being developed include cylindrical blocks for volvo car engines, door panels, floor panels, space frames, truck bodies, armored vehicles, rims of the wheels, engine components, subframes and engine cradles, fuel tanks, frames of motorcycles and bicycles, etc.

Aerospace sectors: In the aerospace sector, R-FSW is used to construct lightweight and high-strength components to enhance the performance and safety of modern aircraft. Components include airframes, fuel tanks, different aircraft parts, rocket components, fuselage panels, skin panels, stringers, wings, parts of airbus, and many more.

Railway sectors: In railway sector, R-FSW is used for fabricating tankers, wagons, container bodies, underground carriages, etc.

Shipbuilding industries: R-FSW has significant applications in the fabrication of marine components due to its ability to create robust, corrosion-resistant joints. In marine industries, it is used to develop marine vessels, deck panels, side panels, bulkheads and floors, pads for helicopter landing, hulls, offshore accommodation and superstructures, etc.

Construction sectors: Components including frames for windows, facade panels, aluminium pipelines, etc., are developed using R-FSW process.

Electrical/Electronics industries: Bus bars, electrical connectors, housings for electric motor, etc.

Renewable energy sectors: R-FSW is used to create reliable and durable joints essential for the long-term performance of wind turbine blades, solar panel frames, tanks, boiler vessels, cylinders, heat exchangers, etc.

4 Conclusions

The present study provides a succinct overview of the fundamental principles, benefits, and diverse applications of mechatronics systems in manufacturing techniques, robotics, and automation. One of the more sophisticated and advanced welding technique, robotic-friction stir welding offers numerous benefits compared to conventional welding techniques. From the current study, the following salient features of the R-FSW process may be highlighted:

(i) The integration of robots, sensors, and automation makes the R-FSW more advantageous.
(ii) Incorporation of a feedback system makes the system more reliable to the users.
(iii) Welding of complex profiles and continuous welding is the main advantage of R-FSW process.
(iv) Wide range of different materials including metal and non-metal can be joined or welded precisely using this process.
(v) Higher accuracy and precision make the process unique and easily adoptable in important sectors including aerospace, marine, defense, etc.
(vi) It is an eco-friendly and cost-efficient method in the manufacturing sector.
(vii) Automation systems facilitate mass production within manufacturing industries.
(viii) Human safety is one of the key elements in industries, and this is very important in this process.

In summary, the current study underscores the multifaceted advantages of R-FSW, ranging from its technical capabilities to its environmental and safety considerations, positioning it as a pivotal technology in modern manufacturing practices.

Potential Areas for Future Research

Ongoing researches are concentrated on augmenting the load capacity of end effectors and its modification, which will enable the welding of thicker materials requiring larger axial forces. This enhancement will open new possibilities for joining heavier and more robust structures, expanding the applicability of R-FSW across various industries. Furthermore, hybrid systems combining different technologies, sensors, and controllers are emerging as promising areas of research. It will enhance productivity, improve weld quality, and enable the fabrication of complex components with greater efficiency. These hybrid systems are expected to streamline manufacturing processes and address the evolving demands of modern industries for innovative and versatile welding solutions.

References

1. Srivastava AK, Kumar N, Dixit AR (2021) Friction stir additive manufacturing—an innovative tool to enhance mechanical and microstructural properties. Mater Sci Eng B 263:114832. https://doi.org/10.1016/J.MSEB.2020.114832
2. Kumar A, Parveen, Liu Y, Kumar R (eds) (2024) Handbook of intelligent and sustainable manufacturing: tools, principles, and strategies, 1st edn. CRC Press. https://doi.org/10.1201/9781003405870
3. Leng J, Wang D, Shen W, Li X, Liu Q, Chen X (2021) Digital twins-based smart manufacturing system design in Industry 4.0: a review. J Manuf Syst 60:119–137. https://doi.org/10.1016/J.JMSY.2021.05.011
4. Iyer S, Sangwan KS, Dhiraj (2024) Development of an industrial symbiosis framework through digitalization in the context of Industry 4.0. Procedia CIRP 122:515–520. https://doi.org/10.1016/J.PROCIR.2024.01.075
5. Ajay, Singh H, Parveen, AlMangour B (eds) (2023) Handbook of smart manufacturing: forecasting the future of Industry 4.0, 1st edn. CRC Press. https://doi.org/10.1201/9781003333760
6. Ajay, Parveen, Kumar A, Mittal RK, Goel R (eds) (2023) Waste recovery and management: an approach toward sustainable development goals, 1st edn. CRC Press. https://doi.org/10.1201/9781003359784
7. Kumar A, Dogra N, Sarita, Bhatia S, Sidhu MS (eds) (2024) Handbook of intelligent and sustainable smart dentistry: nature and bio-inspired approaches, processes, materials, and manufacturing, 1st edn. CRC Press. https://doi.org/10.1201/9781003404934
8. Ahmed MB, Sanin C, Szczerbicki E (2019) Smart virtual product development (SVPD) to enhance product manufacturing in Industry 4.0. Procedia Comput Sci 159:2232–2239. https://doi.org/10.1016/J.PROCS.2019.09.398
9. Kumar A, Kumar P, Dogra N, Jaglan A (2023) Application of incremental sheet forming (ISF) toward biomedical and medical implants. In: Handbook of flexible and smart sheet forming techniques: Industry 4.0 approaches, pp 247–263. https://doi.org/10.1002/9781119986454.ch13
10. Kumar A, Rajak DK, Kumar P, Rathee S (eds) (2024) Smart electric and hybrid vehicles: design, modeling, and assessment by Industry 4.0 approaches, 1st edn. CRC Press. https://doi.org/10.1201/9781003502470
11. Amerongen JV, Breedveld P (2003) Modelling of physical systems for the design and control of mechatronic systems. Annu Rev Control 27(1):87–117. https://doi.org/10.1016/S1367-5788(03)00010-5
12. Shi Y, Zhang K (2021) Advanced model predictive control framework for autonomous intelligent mechatronic systems: a tutorial overview and perspectives. Annu Rev Control 52:170–196. https://doi.org/10.1016/J.ARCONTROL.2021.10.008
13. Lee MH, Nicholls HR (1999) Review article tactile sensing for mechatronics—a state of the art survey. Mechatronics 9(1):1–31. https://doi.org/10.1016/S0957-4158(98)00045-2
14. Perner M, Algermissen S, Keimer R, Monner HP (2016) Avoiding defects in manufacturing processes: a review for automated CFRP production. Robot Comput Integr Manuf 38:82–92. https://doi.org/10.1016/J.RCIM.2015.10.008
15. Gharbia M, Chang-Richards A, Lu Y, Zhong RY, Li H (2020) Robotic technologies for on-site building construction: a systematic review. J Build Eng 32:101584. https://doi.org/10.1016/J.JOBE.2020.101584

16. Kumar A, Rajak DK, Kumar P, Rathee S (eds) (2024) Smart electric and hybrid vehicles: fundamentals, strategies and applications, 1st edn. CRC Press. https://doi.org/10.1201/9781003495574
17. Aly BA, Low T, Long D, Baillie C, Brett P (2023) Robotics and sensing technologies in red meat processing: a review. Trends Food Sci Technol 137:142–155. https://doi.org/10.1016/J.TIFS.2023.05.015
18. Rani S, Tripathi K, Kumar A (2023) Machine learning aided malware detection for secure and smart manufacturing: a comprehensive analysis of the state of the art. Int J Interact Des Manuf (IJIDeM). https://doi.org/10.1007/s12008-023-01578-0
19. Kumar A, Kumar D, Kumar P, Dhawan V (2020) Optimization of incremental sheet forming process using artificial intelligence-based techniques. In: Nature-inspired optimization in advanced manufacturing processes and systems. CRC Press, pp 113–130. https://doi.org/10.1201/9781003081166-8
20. Mohanraj T, Shankar S, Rajasekar R, Sakthivel NR, Pramanik A (2020) Tool condition monitoring techniques in milling process—a review. J Market Res 9(1):1032–1042. https://doi.org/10.1016/J.JMRT.2019.10.031
21. Xu W, He Y, Li J, Zhou J, Xu E, Wang W, Liu D (2023) Robotization and intelligent digital systems in the meat cutting industry: from the perspectives of robotic cutting, perception, and digital development. Trends Food Sci Technol 135:234–251. https://doi.org/10.1016/J.TIFS.2023.03.018
22. Azamfirei V, Psarommatis F, Lagrosen Y (2023) Application of automation for in-line quality inspection, a zero-defect manufacturing approach. J Manuf Syst 67:1–22. https://doi.org/10.1016/J.JMSY.2022.12.010
23. Gao H, Jiao X, Zhou C, Shen Q, Yu Y (2011) Study on remote control underwater welding technology applied in nuclear power station. Procedia Eng 15:4988–4993. https://doi.org/10.1016/J.PROENG.2011.08.927
24. Kadian N, Rathee R, Kumar A (2023) Heat-assisted dieless sheet forming techniques for hard-to-form materials. In: Handbook of sustainable materials: modelling, characterization, and optimization. CRC Press eBooks, pp 457–468. https://doi.org/10.1201/9781003297772-22
25. Kumar A, Kumar P, Srivastava AK, Goyat V (eds) (2023) Modeling, characterization, and processing of smart materials. In: Advances in chemical and materials engineering book series. https://doi.org/10.4018/978-1-6684-9224-6. ISBN13: 9781668492246
26. Ranjan R, Jha SK, Kumar A (2024) Enhancing hardness in SMAW welded joints with low-frequency vibration assistance: an experimental and analytical study. J Adhes Sci Technol 1–25. https://doi.org/10.1080/01694243.2024.2371227
27. Kumar A, Gahlot P, Kumar A, Phanden RK (2024) A state-of-the-art literature review on friction stir welding of 7075-aluminium alloy for tool geometry selection, environmental parameter and mathematical modelling perspective. Int J Interact Des Manuf (IJIDeM). https://doi.org/10.1007/s12008-024-01922-y
28. Chowdhury I, Sengupta K, Chandra PC, Kumar A, Mondal AK (2024) Progressive effect of dual-hybridization in friction stir welding by ultrasonic vibration and resistive heating for joining dissimilar material Al6063 aluminium alloy and C26000 copper alloy. J Adhes Sci Technol 1–21. https://doi.org/10.1080/01694243.2024.2359842
29. Burande DV, Kalita K, Gupta R, Kumar A, Chohan JS, Kumar D (2024) Machine learning metamodels for thermo-mechanical analysis of friction stir welding. Int J Interact Des Manuf (IJIDeM). https://doi.org/10.1007/s12008-024-01871-6
30. Murmu SK, Chattopadhayaya S, Cep R, Kumar A, Kumar A, Mahato SK, Kumar A, Sethy PR, Logesh K (2024) Exploring tribological properties in the design and manufacturing of metal matrix composites: an investigation into the AL6061-SiC-fly ASH alloy fabricated via stir casting process. Front Mater 11. https://doi.org/10.3389/fmats.2024.1415907
31. Saxena A, Gupta TK, Srivastava R, Srivastava AK, Kumar A (2024) A comparison of different experimental approaches in the determination of dynamic fracture toughness (J1d) behavior for RHA steel cost sustainable MMAW weldments. J Adhes Sci Technol 1–22. https://doi.org/10.1080/01694243.2024.2334267

32. Goswami E, Srivastava AK, Kumar A, Salman M, Choudhary MZ, Pandey SS (2023) Review of microstructural and mechanical properties of microwave welding of lightweight alloys. In: Modeling, characterization, and processing of smart materials. IGI Global, pp 185–204. https://doi.org/10.4018/978-1-6684-9224-6.ch009
33. Sen M, Puri AB (2022) Formation of intermetallic compounds (IMCs) in FSW of aluminum and magnesium alloys (Al/Mg alloys)—a review. Mater Today Commun 33:105017. https://doi.org/10.1016/j.mtcomm.2022.105017
34. Ajay, Parveen, Ahmad S, Sharma J, Gambhir V (eds) (2023) Handbook of sustainable materials: modelling, characterization, and optimization, 1st edn. CRC Press. https://doi.org/10.1201/9781003297772
35. Sen M, Shankar S, Chattopadhyaya S (2019) Micro-friction stir welding (μFSW)—a review. Mater Today: Proc 27:2469–2473. https://doi.org/10.1016/j.matpr.2019.09.220
36. Srivastava AK, Tiwari S, Pachauri P, Gupta N, Sunil B, Kumar A (2023) Bonding strength and microstructural features of Al5083-AZ31B alloys laminated sheet through friction stir additive manufacturing. J Adhes Sci Technol 38(4):583–596. https://doi.org/10.1080/01694243.2023.2240637
37. El-Sayed MM, Shash AY, Abd-Rabou M, ElSherbiny MG (2021) Welding and processing of metallic materials by using friction stir technique: a review. J Adv Join Process 3:100059. https://doi.org/10.1016/J.JAJP.2021.100059
38. Huang Y, Meng X, Xie Y, Wan L, Lv Z, Cao J, Feng J (2018) Friction stir welding/processing of polymers and polymer matrix composites. Compos A Appl Sci Manuf 105:235–257. https://doi.org/10.1016/J.COMPOSITESA.2017.12.005
39. Hwang YM, Kang ZW, Chiou YC, Hsu HH (2008) Experimental study on temperature distributions within the workpiece during friction stir welding of aluminum alloys. Int J Mach Tools Manuf 48(7–8):778–787. https://doi.org/10.1016/J.IJMACHTOOLS.2007.12.003
40. Mandal S, Rice J, Elmustafa AA (2008) Experimental and numerical investigation of the plunge stage in friction stir welding. J Mater Process Technol 203(1–3):411–419. https://doi.org/10.1016/J.JMATPROTEC.2007.10.067
41. Okawa Y, Taniguchi M, Sugii H, Marutani Y (2006) Development of 5-axis friction stir welding system. SICE-ICASE Int Joint Conf 2006:1266–1269. https://doi.org/10.1109/SICE.2006.315435
42. Patel V, Wouters H, Baghdadchi A, de Backer J, Igestrand M, Azimi S, Andersson J (2023) Robotic friction stir welding in lightweight battery assembly of extrusion-cast aluminium alloys. J Adv Join Process 8:100156. https://doi.org/10.1016/j.jajp.2023.100156
43. Zhao J, Duan Y, Xie B, Zhang Z (2021) FSW robot system dimensional optimization and trajectory planning based on soft stiffness indices. J Manuf Process 63:88–97. https://doi.org/10.1016/j.jmapro.2020.05.004
44. Soron M, Kalaykov I (2006) A robot prototype for friction stir welding. In: 2006 IEEE conference on robotics, automation and mechatronics, pp 1–5. https://doi.org/10.1109/RAMECH.2006.252646
45. Mendes N, Neto P, Simão MA, Loureiro A, Pires JN (2016) A novel friction stir welding robotic platform: welding polymeric materials. Int J Adv Manuf Technol 85(1–4):37–46. https://doi.org/10.1007/s00170-014-6024-z
46. Mendes N, Neto P, Loureiro A (2014) Robotic friction stir welding aided by hybrid force/motion control. In: Proceedings of the 2014 IEEE emerging technology and factory automation (ETFA), pp 1–4. https://doi.org/10.1109/ETFA.2014.7005266
47. Kashaev N, Ventzke V, Çam G (2018) Prospects of laser beam welding and friction stir welding processes for aluminum airframe structural applications. J Manuf Process 36:571–600. https://doi.org/10.1016/J.JMAPRO.2018.10.005
48. Borse MN, Bauri R, Shankar S (2023) Study of microstructure evolution of friction stir welded novel (Al-Zn-Mg)-Fe (HE700) cast alloys for automotive applications. Mater Sci Eng A 879:145274. https://doi.org/10.1016/J.MSEA.2023.145274
49. di Bella G, Alderucci T, Favaloro F, Borsellino C (2023) Effect of tool tilt angle on mechanical resistance of AA6082/AA5083 friction stir welded joints for marine applications. Procedia CIRP 118:879–884. https://doi.org/10.1016/J.PROCIR.2023.06.151

50. Kumar S, Mahajan A, Kumar S, Singh H (2022) Friction stir welding: types, merits & demerits, applications, process variables & effect of tool pin profile. Mater Today: Proc 56:3051–3057. https://doi.org/10.1016/J.MATPR.2021.12.097
51. Cook GE, Crawford R, Clark DE, Strauss AM (2004) Robotic friction stir welding. Ind Robot 31(1):55–63. https://doi.org/10.1108/01439910410512000
52. Bres A, Monsarrat B, Dubourg L, Birglen L, Perron C, Jahazi M, Baron L (2010) Simulation of friction stir welding using industrial robots. Ind Robot 37(1):36–50. https://doi.org/10.1108/01439911011009948
53. Silva-Magalhães A, Cederqvist L, de Backer J, Håkansson E, Ossiansson B, Bolmsjö G (2019) A friction stir welding case study using temperature controlled robotics with a HPDC cylinder block and dissimilar materials joining. J Manuf Process 46:177–184. https://doi.org/10.1016/j.jmapro.2019.08.012
54. Yavuz H (2004) Function-oriented design of a friction stir welding robot. J Intell Manuf 15:761–775. https://doi.org/10.1023/B:JIMS.0000042662.79454.c5
55. Gibson BT, Lammlein DH, Prater TJ, Longhurst WR, Cox CD, Ballun MC, Dharmaraj KJ, Cook GE, Strauss AM (2014) Friction stir welding: process, automation, and control. J Manuf Process 16(1):56–73. https://doi.org/10.1016/j.jmapro.2013.04.002
56. Wanjara P, Monsarrat B, Larose S (2013) Gap tolerance allowance and robotic operational window for friction stir butt welding of AA6061. J Mater Process Technol 213(4):631–640. https://doi.org/10.1016/j.jmatprotec.2012.10.010
57. Guillo M, Dubourg L (2016) Impact & improvement of tool deviation in friction stir welding: Weld quality & real-time compensation on an industrial robot. Robot Comput Integr Manuf 39:22–31. https://doi.org/10.1016/j.rcim.2015.11.001
58. Wang K, Léonard F, Abba G (2015) Dynamic model identification of axial force in robotic friction stir welding. IFAC-PapersOnLine 28(3):1936–1941. https://doi.org/10.1016/j.ifacol.2015.06.370
59. Gurdal O, Rae B, Zonuzi A, Ozturk E (2019) Vision-assisted robotic finishing of friction stir-welded corner joints. Procedia Manuf 40:70–76. https://doi.org/10.1016/j.promfg.2020.02.013
60. Xiao J, Wang M, Liu H, Liu S, Zhao H, Gao J (2023) A constant plunge depth control strategy for robotic FSW based on online trajectory generation. Robot Comput Integr Manuf 80. https://doi.org/10.1016/j.rcim.2022.102479
61. Yue W, Liu H, Huang T (2022) An approach for predicting stiffness of a 5-DOF hybrid robot for friction stir welding. Mech Mach Theory 175. https://doi.org/10.1016/j.mechmachtheory.2022.104941
62. Zhao X, Kalya P, Landers RG, Krishnamurthy K (2007) Design and implementation of a nonlinear axial force controller for friction stir welding processes. Am Control Conf 2007:5553–5558. https://doi.org/10.1109/ACC.2007.4282731
63. Grimm A, Schulze S, Silva A, Göbel G, Standfuss J, Brenner B, Beyer E, Füssel U (2015) Friction stir welding of light metals for industrial applications. Mater Today: Proc 2:S169–S178. https://doi.org/10.1016/j.matpr.2015.05.007
64. Li Q, Wu W, Xiang J, Li H, Wu C (2015) A hybrid robot for friction stir welding. Proc Inst Mech Eng C J Mech Eng Sci 229(14):2639–2650. https://doi.org/10.1177/0954406214562848
65. Li Z, Zhao H, Zhang X, Dong J, Hu L, Gao H (2023) Multi-parameter sensing of robotic friction stir welding based on laser circular scanning. J Manuf Process 89:92–104. https://doi.org/10.1016/j.jmapro.2023.01.060
66. Busu N, Jaffarullah MS, Low CY, Shaari MSB, Armansyah, Jaffar A (2015) A review of force control techniques in friction stir process. Procedia Comput Sci 76:528–533. https://doi.org/10.1016/J.PROCS.2015.12.331
67. Blignault C, Hattingh DG, Kruger GH, van Niekerk TI, James MN (2008) Friction stir weld process evaluation by multi-axial transducer. Measurement 41(1):32–43. https://doi.org/10.1016/j.measurement.2006.12.001
68. Mendes N, Neto P, Loureiro A, Moreira AP (2016) Machines and control systems for friction stir welding: a review. Mater Des 90:256–265. https://doi.org/10.1016/j.matdes.2015.10.124

69. Gibson BT, Cox CD, Longhurst WR, Strauss AM, Cook GE (2012) Exploiting robotic link deflection for low-cost force measurement in manufacturing. Measurement 45(1):140–143. https://doi.org/10.1016/j.measurement.2011.09.012
70. Lammlein DH, Gibson BT, Delapp DR, Cox C, Strauss AM, Cook GE (2012) The friction stir welding of small-diameter pipe: an experimental and numerical proof of concept for automation and manufacturing. Proc Inst Mech Eng Part B: J Eng Manuf 226(3):383–398. https://doi.org/10.1177/0954405411402767
71. Longhurst WR, Strauss AM, Cook GE, Cox CD, Hendricks CE, Gibson BT, Dawant YS (2010) Investigation of force-controlled friction stir welding for manufacturing and automation. Proc Inst Mech Eng Part B: J Eng Manuf 224(6):937–949. https://doi.org/10.1243/09544054JEM1709
72. Mendes N, Loureiro A, Martins C, Neto P, Pires JN (2014) Morphology and strength of acrylonitrile butadiene styrene welds performed by robotic friction stir welding. Mater Des 64:81–90. https://doi.org/10.1016/j.matdes.2014.07.047
73. Yang C, Youcef-Toumi K (2022) Principle, implementation, and applications of charge control for piezo-actuated nanopositioners: a comprehensive review. Mech Syst Signal Process 171:108885. https://doi.org/10.1016/j.ymssp.2022.108885
74. Longhurst WR, Strauss AM, Cook GE, Fleming PA (2010) Torque control of friction stir welding for manufacturing and automation. Int J Adv Manuf Technol 51:905–913. https://doi.org/10.1007/s00170-010-2678-3
75. Barbieri G, Fantuzzi C, Borsari R (2014) A model-based design methodology for the development of mechatronic systems. Mechatronics 24(7):833–843. https://doi.org/10.1016/j.mechatronics.2013.12.004
76. Quan L, Zhao W (2024) A review on positioning uncertainty in motion control for machine tool feed drives. Precis Eng 88:428–448. https://doi.org/10.1016/j.precisioneng.2024.03.003
77. Mendes N, Neto P, Pires JN, Loureiro A (2013) An optimal fuzzy-PI force/motion controller to increase industrial robot autonomy. Int J Adv Manuf Technol 68:435–441. https://doi.org/10.1007/s00170-013-4741-3
78. Thomas WM, Johnson KI, Wiesner CS (2003) Friction stir welding-recent developments in tool and process technologies. Adv Eng Mater 5(7):485–490. https://doi.org/10.1002/adem.200300355
79. Hayati S, Rankin A, Kim W, Leger P, Castano R, Ali K (2007) Advanced robotics technology infusion to the Nasa Mars exploration rover (MER) Project. IFAC Proc Vol 40(15):180–185. https://doi.org/10.3182/20070903-3-FR-2921.00033
80. Srivastava AK, Kumar A, Kumar P, Gautam P, Dogra N (2023) Research progress in metal additive manufacturing: challenges and opportunities. Int J Interact Des Manuf (IJIDeM). https://doi.org/10.1007/s12008-023-01661-6
81. Ku N, Ha S, Roh M (2014) Design of controller for mobile robot in welding process of shipbuilding engineering. J Comput Des Eng 1(4):243–255. https://doi.org/10.7315/JCDE.2014.024
82. Zhang HW, Zhang Z, Chen JT (2005) The finite element simulation of the friction stir welding process. Mater Sci Eng A 403(1–2):340–348. https://doi.org/10.1016/j.msea.2005.05.052
83. Rajakumar S, Muralidharan C, Balasubramanian V (2011) Influence of friction stir welding process and tool parameters on strength properties of AA7075-T6 aluminium alloy joints. Mater Des 32(2):535–549. https://doi.org/10.1016/j.matdes.2010.08.025
84. Meng X, Huang Y, Cao J, Shen J, dos Santos JF (2021) Recent progress on control strategies for inherent issues in friction stir welding. Prog Mater Sci 115:100706. https://doi.org/10.1016/j.pmatsci.2020.100706
85. Wahidi SI, Oterkus S, Oterkus E (2024) Robotic welding techniques in marine structures and production processes: a systematic literature review. Mar Struct 95:103608. https://doi.org/10.1016/j.marstruc.2024.103608
86. Fleming PA, Hendricks CE, Wilkes DM, Cook GE, Strauss AM (2009) Automatic seam-tracking of friction stir welded T-joints. Int J Adv Manuf Technol 45:490–495. https://doi.org/10.1007/s00170-009-1990-2

87. Shi J, Wang Y, Zhang G, Ding H (2013) Optimal design of 3-DOF PKM module for friction stir welding. Int J Adv Manuf Technol 66:1879–1889. https://doi.org/10.1007/s00170-012-4467-7
88. Longhurst WR, Cox CD, Gibson BT, Cook GE, Strauss AM, Wilbur IC, Osborne BE (2017) Development of friction stir welding technologies for in-space manufacturing. Int J Adv Manuf Technol 90:81–91. https://doi.org/10.1007/s00170-016-9362-1
89. Buffa G, Fratini L, Ruisi V (2009) Friction stir welding of tailored joints for industrial applications. IntJ Mater Form 2(Suppl 1):311–314. https://doi.org/10.1007/s12289-009-0579-5
90. Shultz EF, Fehrenbacher A, Pfefferkorn FE, Zinn MR, Ferrier NJ (2013) Shared control of robotic friction stir welding in the presence of imperfect joint fit-up. J Manuf Process 15(1):25–33. https://doi.org/10.1016/j.jmapro.2012.07.002
91. Bi ZM, Miao Z, Zhang B, Zhang CWJ (2020) The state of the art of testing standards for integrated robotic systems. Robot Comput Integr Manuf 63:101893. https://doi.org/10.1016/j.rcim.2019.101893
92. Natarajan M, Pasupuleti T, Giri J, Al-Lohedan HA, Katta LN, Mohammad F, Sunheriya N, Chadge R, Mahatme C, Giri P, Mallik S, Sathish T (2024) Optimization of wire spark erosion machining of Grade 9 titanium alloy (Grade 9) using a hybrid learning algorithm. AIP Adv 14(1). https://doi.org/10.1063/5.0177658
93. Narasimhamu KL, Natarajan M, Thejasree P, Makki E, Giri J, Sunheriya N, Chadge R, Mahatme C, Giri P, Sathish T (2024) Development of hybrid optimization model using grey-ANFIS-Jaya algorithm for CNC drilling of aluminium alloy. J Eng 2024:1–12. https://doi.org/10.1155/2024/1476770
94. Tufail MS, Giri J, Makki E, Sathish T, Chadge R, Sunheriya N (2024) Machinability of different cutting tool materials for electric discharge machining: a review and future prospects. AIP Adv 14(4). https://doi.org/10.1063/5.0201614
95. Giri J, Sunheriya N, Sathish T, Kadu Y, Chadge R, Giri P, Parthiban A, Mahatme C (2024) Optimization of process parameters to improve mechanical properties of fused deposition method using Taguchi method. Interactions 245(1). https://doi.org/10.1007/s10751-024-01925-x
96. Kamble PD, Giri J, Makki E, Sunheriya N, Sahare SB, Chadge R, Mahatme C, Giri P, Panchal H (2024) An application of hybrid Taguchi-ANN to predict tool wear for turning EN24 material. AIP Adv 14(1). https://doi.org/10.1063/5.0186432

Comparative Analysis of OFDM and FBMC System Using Cognitive Radio Technique

Rupayali Swaroop, Dinesh Sethi, and Girraj Sharma

Abstract In mobile communication, spectrum range is a limitation as an enormous amount of bandwidth is not appropriately utilized which reduces the Quality of Services (QoS). The spectral efficacy can be increased if the spectrum is accessible without interfering. Though OFDM has several advantages for removing the ISI (Inter Symbol Interference) from the channel, Cyclic Prefix (CP) is introduced; however, it occupies excess bandwidth. To eliminate these issues, FBMC is an appropriate applicant as it comprises a bank of filters and it also supports MIMO (Multiple Input Multiple Output) systems. The proposed work focuses on integrating the energy detection schemes with OFDM and FBMC waveforms. The study compares the simulations and comparative analysis of FBMC and OFDM with different aspects to analyze spectrum sensing. The projected work is studied thoroughly and implemented on MATLAB version 2019. Several parameters such as Probability of Detection (Pd), Probability of False Alarm (PFA), and Bit Error Rate (BER) are estimated for OFDM and FBMC systems. It is seen that the proposed spectrum sensing method works more efficiently in FBMC as compared with the OFDM system. The methods used in FBMC deliver compatibility with MIMO. It also decreases PAPR with minimum BER. These advantages make FBMC a suitable applicant for 5G systems and also used for IoT applications.

Keywords Spectrum sensing · OFDM · FBMC · Cognitive radio · Energy detection

R. Swaroop (✉) · D. Sethi · G. Sharma
Electronics and Communication Department, JECRC University, Jaipur, India
e-mail: rupayali.swaroop@jecrcu.edu.in

D. Sethi
e-mail: dinesh.sethi@redffmail.com

G. Sharma
e-mail: girrajsharma.ece@jecrc.ac.in

A. Kumar et al. (eds.), *Advances in Mechatronics Systems*, Emerging Trends in Mechatronics, https://doi.org/10.1007/978-981-97-8472-1_6

1 Introduction

The constant development of wireless devices creates a cluster in ISM (Industry, Scientific, and Medical) bands due to which the spectrum is not properly utilized to its optimal level. Nowadays, a smart technique is used i.e. Cognitive radio (Cr) which automatically detects the ideal spectrum for licensed and unlicensed users. If the spectrum is idle then the unlicensed user can use the vacant spectrum without interference but if a licensed user is detected by the Cr [1]. It will switch to another alternative route for unlicensed users. By using this technology the spectrum can be utilized efficiently.

The Federal Communications Commission (FCC) in 2008 issued a rule that allows using the unlicensed user in the TV broadcast spectrum without any interference to licensed users which leads to the development of Cr. This intelligent technology leads to the efficient usage of the spectrum and uses the dynamic bandwidth approach [2]. At the time of spectrum congestion, the Cr permits the optimum usage of the spectral bands that are unutilized by the licensed users. This smart scheme accesses the availability of the licensed user and if it is present then it starts sensing the white space bandwidth for transmitting the data of secondary users without interrupting the primary users. Cr technique [3] can be divided into three types: Energy Detection scheme, Matched Filter scheme, and Cyclostationary scheme. The Energy Detection (ED) [4] method is the basic and simplest system for sensing the signal's absence and existence. The major issue allied with ED is that it delivers high performance merely with high SNR. At low SNR, it provides poor performance. Additionally, the challenge is channel reception quality expectations in Cr. The unlicensed users have access to the licensed channel but without interference. OFDM is the latest transmission scheme utilized in many applications such as LTE (Long-Term Evolution), WiMAX (Worldwide Interoperability for Microwave Access), LTE A (Long-Term Evolution—Advanced), DSL (Digital subscriber line), and so on [4–6]. The basic concept in the waveform technique is distributing one carrier into a number of multi-sub carriers. Each sub-carrier has a narrow carrier which is compared to a particular carrier scheme. At the transmitter, each data symbol [7] gets modulated by each stream and demodulation is applied at the receiver side of the system. In the OFDM [8] module the main problem is the wastage of the spectrum. To minimize this Inter Symbol Interference (ISI), Cyclic Prefix (CP) is inserted between symbols. Researchers are in the process of developing a technique to overcome the wastage of bandwidth in the OFDM scheme. However, ED is considered to be the required technique for overcoming the problem faced by the OFDM scheme and also the use of the vacant bandwidth. In this proposed chapter, the OFDM technique is used with the presence and absence of CP; a filter is added at the transceiver and integrated with the ED, to analyze the efficacy of the system [9, 10].

A technique is proposed in [11], the M-estimator-based channel estimation scheme which is used to compare linear minimum mean-squared error (LMMSE) with conventional least squares (LS) to improve the QOS in the system. The channel is compared with a Rician and Rayleigh fading channel. The better gain performance is achieved in this paper using the M estimator scheme. A Single extension Windowing (SW) is proposed in [12], which is used to reduce OOB (Out-Of-Band) power to improve the performance of the CP OFDM system. A gradual waveform technique is applied which can accurately and effectively reduce the OOB power. This paper then compares conventional CP OFDM with windowing CP OFDM. In [13] Helmert-SLM (Selected mapping) is proposed which is used to reduce PAPR. On comparing SLM OFDM with the Riemann sequence, it provides an improvement of 3.5 dB while using SLM-OFDM with cyclic Hadamard sequence, it provides 4.5 dB. This results in extremely low PAPR and better sensing probability. A memory polynomial technique in [14] proposes a Recursive Least Square (RLS) algorithm that reduces the BER by 27.778%, demodulated OFDM signal reaches the destination without deviating, and Error Vector Magnitude (EVM) by 67.961% at the SNR of 8 dB. This adaptive technique improves the bandwidth utilization and performance of the network.

Advanced multi-carrier schemes such as FBMC, UFMC, NOMA, and so on are promising applicants for upcoming cellular systems. FBMC is the most efficient scheme for 5G mobile communication as it offers high-speed data, minimal interference, increased spectrum efficacy, and better bandwidth usage as compared with other multi-carrier schemes. FBMC system originated from OFDM; it focuses on overcoming the problems faced by the OFDM system. Moreover, CP is not used in this scheme but a bank of filters is applied at the transceiver for better efficacy. The guard interval is used for preventing the ISI (Inter Symbol Interference) though FBMC is not dependent on the guard interval and uses a filter bank for preventing the ISI and ICI (Intercarrier Interference). The merger of modulation and multiplexing by distributing the wide-band channel into a number of narrow-band channels is known as a sub-channel. The ability to solve the possible errors of synchronization makes this scheme more efficient. FBMC technique is proposed in [15] which uses the algorithm of particle swarm optimization (PSO) that reduces the PAPR at the operating channel. The results are verified at different parameters that will provide a reduction in PAPR with less complexity. A comparative study is proposed in [16] comparing filtered OFDM and FBMC. The total analysis of data consists of different parameters like BER performance, complexity of the system, and latency. The inference introduced between the subcarriers is a major problem in CR. Conducting experiments with different parameters and constraints shows that FBMC has better results. Hence, FBMC is a better option for future wireless systems. An optimization algorithm proposed in [17] provides the real and dynamic time programmability configurations at the operating frequency channel. The FBMC system has been compared with various parameters and thus, shows that FBMC performance in throughput and interference is better than the OFDM system in CR.

FBMC has the potential to be an essential future technology for 5G and beyond, enabling higher data speeds, enhanced connectivity in challenging environments, and higher QoS. It depends on several factors, including standardization, compatibility with existing systems, and the development of cost-effective hardware implementations. Additionally, it has the ability to facilitate low-latency communication which is valuable for new applications such as real-time gaming and the IoT.

In this chapter, Sect. 1 comprises the overview of OFDM and FBMC. In Sect. 2, methodology of OFDM and FBMC has been discussed. Section 3, presents the results obtained while comparing the two systems with different parameters, and Sect. 4 will outline as the conclusion.

2 Methodology

The proposed framework is executed using the software MATLAB version 2019. The implemented variables and parameters are described briefly.

2.1 OFDM System Model

In Fig. 1, the random data is produced via a Bernoulli generator (generates binary numbers i.e. 0 and 1) and the transmitter, of OFDM, transforms through Inverse Fast Fourier transform (IFFT). Therefore the signals are converted from the frequency-domain to the time-domain. Here to reduce the ISI [18], another thing CP is placed between the signals of OFDM. These signals are propagated over additive white Gaussian noise (AWGN). Once the signals are received then FFT and protector time intervals are detached and as a result demodulation process is implemented, which leads to the conversion of frequency signals to time domain signals [19].

Let us consider Z, to be the transmitting signal as shown below [20],

$$Z = (z1 + z2, ..., Z(n)) \tag{1}$$

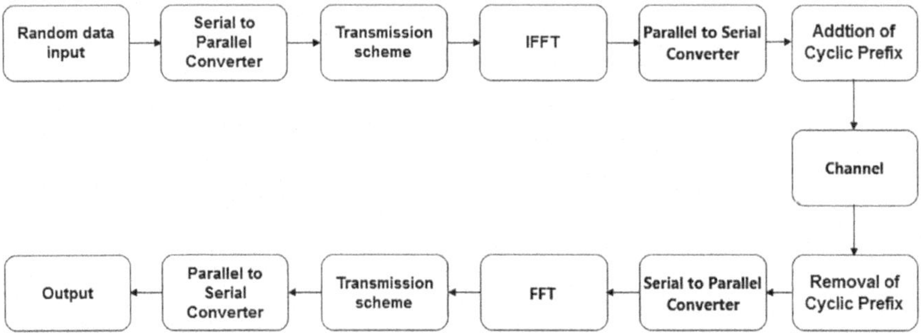

Fig. 1 OFDM transceiver block diagram

The modulated signal of IFFT using Eq. (1) is shown [21]:

$$Z(k) = \frac{\sum_{n=0}^{N-1} X(n)e^{\frac{-j2\pi Kn}{n}}}{\sqrt{N_0}\left(\frac{1}{n}\sum_{k=0}^{N-1}\sum_{n=0}^{N-1} X(n)e^{\frac{j2\pi Kn}{N}} * \sum_{k=0}^{N-1} X(n)e^{\frac{-j2\pi Kn}{N}}\right)} \tag{2}$$

where N_0 defines the number of data/information sub-carriers; and the N defines the m-number of trails. As shown below the CP is inserted into each of the symbols which is used to minimize the ISI.

$$[Zc(K) = Z(K)(E - T + 1 : E)Z(K)] \tag{3}$$

where, E is the $Z(k)$ symbol end sequence and T is total CP.

$$C = \sum_{i=0}^{L-1'} a_i e^{-j2\pi fcp\tau i} \tag{4}$$

where C is the multipath-fading coefficient, $L =$ total paths, $\tau i =$ delay and $a_i =$ attenuation. In the parliamentary- law which is used to thin out noise an equalizer is applied. The feedback of multipath propagated signal is then calculated and equalized at the receiver side is shown below:

$$C(f) = \left(\sum_{n=0}^{N-1} Ce^{\frac{-j2\pi Kn}{n}}\right)\left(\frac{1}{n}\sum_{k=0}^{N-1}\sum_{n=0}^{N-1} Ce^{\frac{j2\pi Kn}{N}} * \sum_{k=0}^{N-1} Ce^{\frac{j2\pi Kn}{N}}\right) \tag{5}$$

where, $C(f)$ is the scheme model of the random noise channel. Now, convoluting the random noise channel and transmitting signal symbol as follows:

$$Xh(t) = Zc(k) * C(f) \tag{6}$$

Here, Gaussian-noise; with variance ($\sigma = 1$) and mean ($\mu = 0$) can be shown by:

$$N = \frac{1}{\sqrt{2}} XT + PN_0T \tag{7}$$

where, N_0T is the total subcarrier when CP is added and P, L is the length of the offset frequency of the carrier.

The AWGN channel with the received data signal can be shown as [21]:

$$Y = C(f)Zc(k) + N \tag{8}$$

After adding CP the energy that is wasted is accounted as $\sqrt{(N_0T/nFFT)}$, using Eq. (8) shown as:

$$Yr = \sqrt{\frac{N_0 T}{nFFT}} * Yh(t) + 10^{\left(-\frac{E/No}{20}\right)*Nt} \tag{9}$$

where, No is the noise and R is the energy received. For noiseless wireless communication, the noise should be less than the energy received. After adding P (carrier offset frequency) in the received signal the equation will be:

$$Y = \left(e^{L*2*\pi*P*\left(Y_0 - \frac{L(Y)-1}{nFFT}\right)}\right) * Y \tag{10}$$

where, $(L(Y)-1)$ is (the total length of the signal received – 1) and Y_0 relates to the zero symbols; of the received signal. After eliminating CP from the receiving signal is shown as: [4]

$$Y = y(T+1:E) \tag{11}$$

Now, FFT is applied at the received signal is shown as:

$$M = \left(\frac{No}{\sqrt{\sum_{n=0}^{N-1} Ye^{\frac{-j2\pi kn}{N}}}}\right) * \left(\sum_{n=0}^{N-1} Y - 1\right) * \left(\sum_{n=0}^{N-1} Ye^{\frac{-j2\pi kn}{N} * \sum_{n=0}^{N-1} Ye^{\frac{j2\pi kn}{N}}}\right) \tag{12}$$

To remove the interference in the receiving signal equalizer is applied as shown as:

$$M = \frac{M}{Cf} \tag{13}$$

BER is shown as:

$$\text{BER} = \frac{\text{Received errors in signal}}{\text{Total transmitted signal}} \tag{14}$$

where, the BER is defined by taking the ratio of errors present in the received signal by total signal transmitted.

2.2 FBMC System Model

Filter Band Multi-Carrier (FBMC) is an advanced waveform technique, which plays a significant role in mobile communication schemes and CR uses. To elaborate further a single-carrier scheme allows its entire spectrum to a single channel, as a result, it develops inter-symbol interferences (ISI) due to which that channel spectrum becomes less than the coherence spectrum. There is another system called Multi-Carrier System (MCS), in which the spectrum bandwidth is distributed among the sub-carriers and these are orthogonal with one another. As concluded, to reduce ISI in this MCS, the spectrum of channels should be less than the spectrum of coherent.

In the year 1971, the technique of multi-carrier system combination of signals was generated through IFFT and the recovery of signals was through FFT, also recognized as OFDM i.e. Orthogonal Frequency Division Multiplexing. In the present times of 4G mobile technology, this OFDM is used as it has an advantage by which the data rate becomes higher and ISI is reduced. The uniqueness of OFDM is the employment of cyclic prefixes between symbols as a result there is no merger or hindrance of symbols with one another, however with the inclusion of CP; this results in the wastage of spectrum. The high PAPR in the OFDM structure is considered to be a disadvantage as it reduces the amplifier's power significantly. Therefore there is a sweeping shift toward 5G mobile technology as it uses FBMC radio technology which uses the spectrum more proficiently without CP [22].

In Fig. 2, the random data is produced via a Bernoulli generator (generates binary numbers i.e. 0 and 1) and the transmitter, of FBMC, processes these signals through the Filters. Therefore the signals are converted to series to parallel converter. The filters are designed to reduce the ISI, minimizing side lobes leakage and enhancing the efficacy of the spectrum. These signals are propagated over additive white Gaussian noise (AWGN). Once the signals are received the demodulation process is implemented, which leads to filtering and then converting the signals to parallel to series conversion which provides the desired output.

FBMC employs a PHYDYAS filter at the modulation and the demodulation. The length of the filter is F, (F = K*N). Now, Discrete Time FBMC's signal can be implemented as.

$$X(n) = \sum_{L=0}^{T-1} \sum_{m \in y}^{l} a_L, mg\left[n - \frac{mT}{2}\right] e^{\frac{j2\pi}{T} l\left(n - \frac{d}{2}\right)} e^{i\varphi k,m}, \tag{15}$$

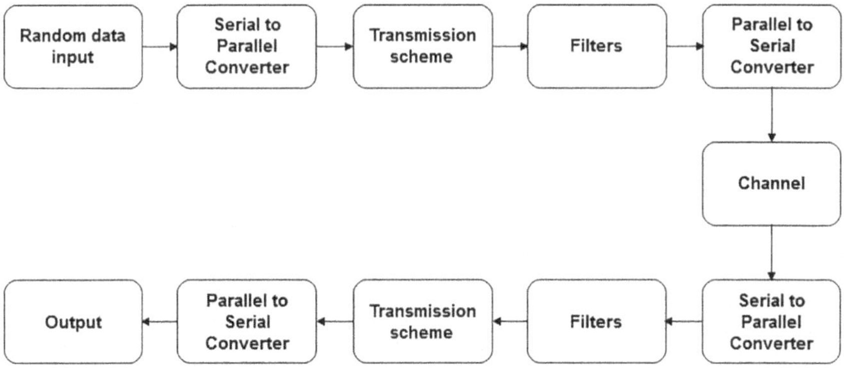

Fig. 2 FBMC transceiver block diagram

where, B = delay, P = total paths T = total sub-carriers, $\varphi_{k,m}$ = phase period, p (n) = prototype-filter and $r_{L,m}$ = real-time symbols transmitted, $e^{i\varphi k,m}$ = is implemented to ensure that between the two input data, there is a phase that is changed by $\frac{\pi}{2}$.

The phase term $\varphi_{L,m}$ is determined for complex signals as shown [23]:

$$\varphi_{k,m} = \varphi_0 + \frac{\pi}{2}(m + L) - \pi mL \tag{16}$$

When $\varphi_0 = 0$, the Eq. (15) can be written as:

$$X(n) = \sum_{L=0}^{N-1} \sum_{m \in Z} r_{L,m} p_{L,m}[n] \tag{17}$$

where, $p_{L,m}[n]$ is the shifted version of the $p[n]$ at frequency and time domain.

Now, demodulated-symbol is can be shown as [23]:

$$A_{L',m'} = s, p_{L',m'} = \sum_{n=-\infty}^{+\infty} x[n] p^*_{L',m'}[n], \tag{18}$$

where, $x[n]$ = real time transmitting symbols and $\sum_{L=0}^{N-1} \sum_{m \in Z} r_{L,m} p_{L,m}[n]$.

Now, Eq. (18) will be re-written as:

$$A_{L',m'} = \sum_{n=-\infty}^{+\infty} \sum_{L=0}^{N-1} \sum_{m \in Z} r_{L,m} p_{L,m}[n] p^*_{L',m'}[n] \tag{19}$$

For real-time orthogonal systems for transmitting and receiving antenna's impulse is shown as [23]:

$$A_{L',m'} = \sum_{n=-\infty}^{+\infty} p_{L0,n0}[n] p^*_{L',m'}[n] \tag{20}$$

$$A_{L',m'} = \sum_{n=-\infty}^{+\infty} p\left[n - \frac{m0T}{2}\right] p\left[n - \frac{mT}{2}\right] e^{\frac{j2\pi}{N}} \left((L_0 - L') . \left(n - \frac{B}{2}\right) e^{i(\varphi_{L,m_0} - \varphi_{L,m'})} \right) \tag{21}$$

By applying, $\Delta k = k - k_0$, $n + m_0 N/2$ and $= m - m_0$,, we can say that:

$$A_{L',m'} = \sum_{n=-\infty}^{+\infty} p[n] p\left[p - \frac{\Delta nT}{2}\right] e^{\frac{j2\pi}{N}} \Delta L \left(\frac{B}{2} - n\right) \cdot e^{i\pi(\Delta L + L0)\Delta m} e^{-i\pi(\Delta L + \Delta m)} \Big) \tag{22}$$

The required received signal $\left(A_{L',m'}\right)$ is achieved by correlating with $e^{-i\pi(\Delta L+\Delta m)}$ and the demodulated signal.

2.3 ED (Energy Detection) Model

The spectral sensing scheme plays an important aspect in the CR cycle. The mainly decides between the two conditions: the licensed user is absent; denoted by H0 and the other condition is denoted by H1 which denotes licensed user exists. The two conditions are written as [24]:

$$H_0 : y(n) = c(n) \tag{23}$$

$$H_1 : y(n) = x(n) + c(n) \tag{24}$$

where, y(n) = signal received, x(n) = signal is transmitted, and c(n) = noise. The descriptive analysis of ED and the noise estimation algorithm is stated in the following.

The Fig. 3, the ED computes that the energy received by N samples are squared-magnitude of Fast Fourier Transform (FFT) are sampled and averaged the values over the N samples, we use the formula as given below [25]:

$$T = \sum_{n=1}^{N} (Y[n])^2 \tag{25}$$

To acquire the detection result, compare the energy, T with a pre-defined threshold value, λ_D;

$T < \lambda_D$ LU signal absent
$T > \lambda_D$ LU signal present.

The performance of the detection technique can be calculated by the probability of detection P_D; and the probability of false alarm P_{FA}.The P_D states the number of accurate detections (Licensed User exists) over to total detecting trails i.e. probability of false alarm states that numeral of times Licensed User is wrongly sensed over total numeral of trials [25]. The possibilities are shown as:

$$P_D = \Pr(T > \lambda_D); H_0 \tag{26}$$

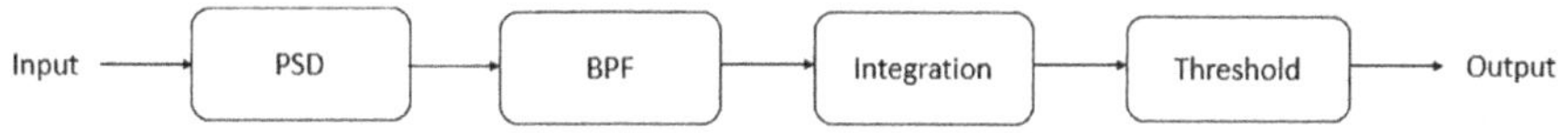

Fig. 3 Energy detection block diagram

$$P_{FA} = \Pr(T > \lambda_D); H_1 \tag{27}$$

where, T relates to energy of N number of samples shown by the Eq. (25) and λ_D; denotes the detecting threshold.

As referred to [26], the false alarm P_{fa} and the possibilities of detection P_d are shown as:

$$P_d = Q\left(\lambda - \frac{\mathrm{N}(\sigma_c^2 + \sigma_x^2)}{(\sigma_c^2 + \sigma_x^2)\sqrt{2N}}\right) \tag{28}$$

$$P_{fa} = Q\left(\frac{\lambda_D - \mathrm{N}\sigma_c^2}{(\sigma_c^2 + \sigma_x^2)\sqrt{2N}}\right) \tag{29}$$

where, Q function, $Q(x) = \frac{1}{2\pi}\int_x^\infty \exp\left(\frac{-u^2}{2}\right)du$, σ_c and σ_s are referred as the standard-deviation of noise and licensed user signal, and N = numeral of trails.

From the Eq. (29) and referring [27], the authors offers the formulation of the sensing threshold λ_D; which can be obtained and shown as:

$$\lambda_D = \sigma_c^2\left(Q^{-1}(P_{fa})\sqrt{2N} + N\right) \tag{30}$$

where $Q^{-1}(.)$ = inverse of Q function.

The threshold calculated depends on parameters such as number of samples, probability of false alarm, and the noise-variance. Hence, prior information of noise is required to obtain a precise threshold [28, 29].

3 Results

Results are studied thoroughly with using the mathematical models of OFDM cognitive schemes and employing it in the MATLAB.

Comparing BER versus SNR

In this section, we are evaluating the performances of OFDM and FBMC with the presence and absence of CP using MATLAB 2018 simulations. In Fig. 4, we are comparing SNR versus BER for OFDM and FBMC with the presence and absence of CP. The preliminary performance approximately starts between 0 to 3 dB. But with the increase in SNR, there is a decrease in BER. The performance of FBMC without CP is giving the desired output. FBMC doesn't require CP and the simulated performances prove that FBMC with no CP is much more reliable and efficient with low ISI as compared to others.

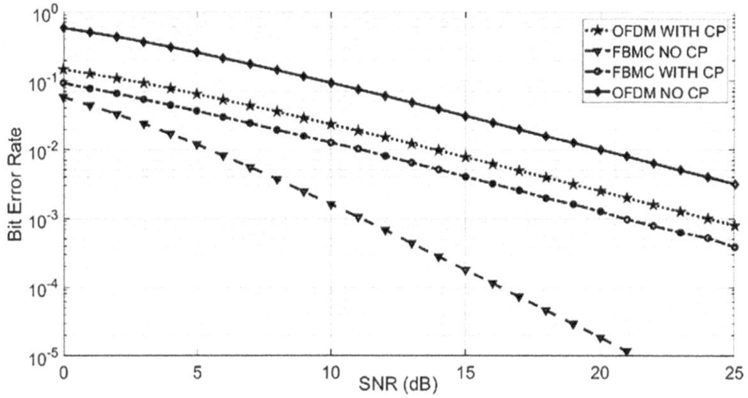

Fig. 4 Comparing BER versus SNR

Comparing Pd versus SNR

In Fig. 5, we are evaluating the SNR versus Probability of energy detection for both OFDM and FBMC with the presence and absence of CP. As we increase the Pd, the SNR is also increased while when Pd is decreased the SNR is also decreased. This indicates that for efficient and reliable spectrum detection, an algorithm of power control is required and the noise power should be less than the received power. Through the graph, we can that FBMC without CP is providing better results.

Comparing Pd versus PFA

In Fig. 6, we are evaluating the Probability of False Alarm versus Probability of energy detection for both OFDM and FBMC with the presence and absence of CP. The Pd is represented as a function for the total numeral of trails, which are mapped

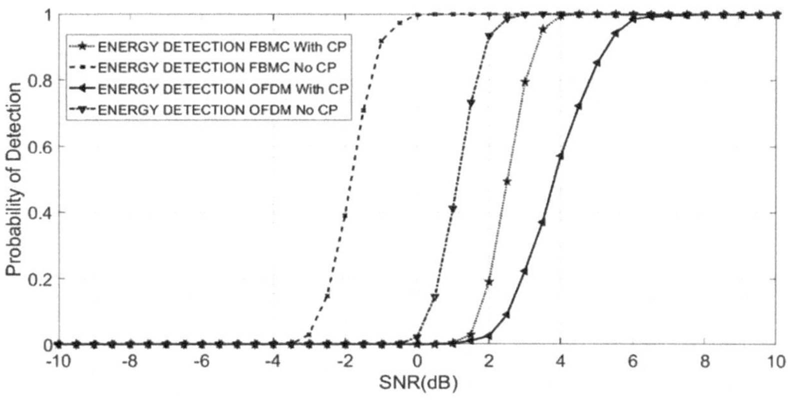

Fig. 5 Comparing Pd versus SNR

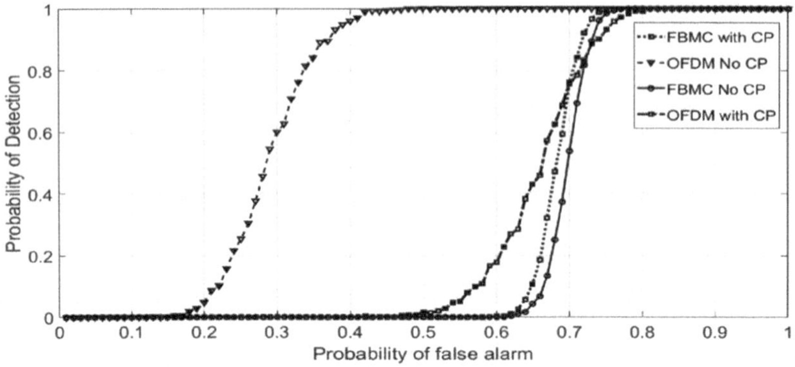

Fig. 6 Comparing Pd versus PFA

on the graph for various values of SNR. From the graph, we can say that, when there is a rise in the number of trails, the Pd increases for high SNR and when the SNR is low the Pd starts decreasing. For analyzing PAPR, identical transmitted signals are used for OFDM and FBMC. The technique used is the Selective mapping (SLM) PAPR reduction scheme. The resulting output shows that FBMC can receive and transmit signals with more efficiency.

Comparing CCDF(Pr[PAPR > PAPRo]) versus PAPRo

In Fig. 7, CCDF of PAPR as a function of Eb/No is performed. After applying PAPR reduction techniques PAPR is reduced. It can be observed that at peak power is minimized to an optimum level. Each PAPR is successfully reducing power consumption. Hence, the performance of FBMC in the absence of CP is better than other reduction schemes.

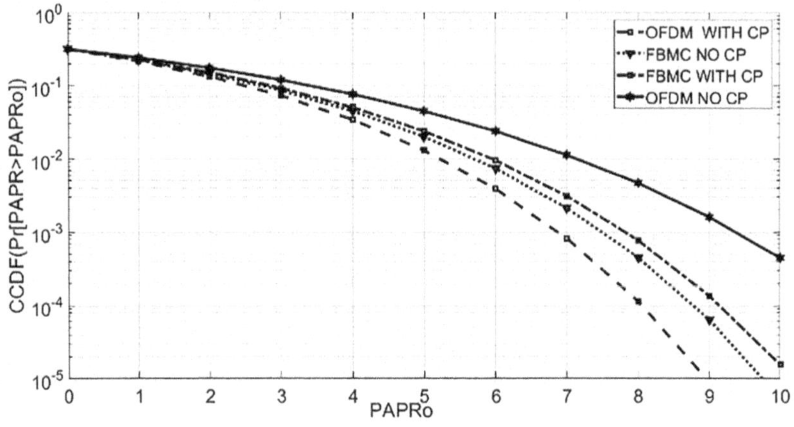

Fig. 7 Comparing CCDF (Pr [PAPR > PAPRo]) versus PAPRo

4 Conclusion

The OFDM system is inadequate for a 5G communication system due to its several shortcomings such as sensitivity to real-time, occurrence of phase distortion, presence of CP, and other factors. For advanced radio, we have analyzed and simulated the FBMC waveform in this work. FBMC comprises a bank of filters that enables minimum sidelobes that prevent ISI for better efficiency and also upgrade the performance of spectrum sensing by reducing the spectrum leakage. The simulations and comparative analysis of FBMC and OFDM are performed with different aspects to analyze spectrum sensing, Peak to Average Power Ratio (PAPR), Probability of Detection (PD), BER measurements, and Probability of False Alarm (PFA). The work indicates that FBMC without CP is performing better than OFDM. The spectrum utilization is more efficient in FBMC as compared with the OFDM system. The FBMC waveform technique is the best wireless communication technique for 5G and it can be efficiently used in IoT appliances.

References

1. Haque TU, Hossen S (2019) PAPR reduction in OFDM system using clipping and filtering methods based on CCDF. Int J Comput Netw Inf Secur 9(10):12
2. Yang J, Song Z, Gao Y, Gu X, Feng Z (2021) Adaptive compressed spectrum sensing for multiband signals. IEEE Trans Wireless Commun 20(11):7642–7654
3. Ijiga OE, Ogundile OO, Familua AD, Versfeld DJ (2019) Review of channel estimation for candidate waveforms of next generation networks. Electronics 8(9):956
4. Swaroop R, Kumar A (2020) A brief study and analysis of NOMA techniques for 5G. In: IEEE international women in engineering (WIE) conference on electrical and computer engineering (WIECON-ECE), pp 13–16
5. Rani S, Kumar A, Bagchi A, Yadav S, Kumar S (2021) RPL based routing protocols for load balancing in IoT network. J Phys: Conf Ser 1950(1):012073
6. Osman AM, Rahman SA, Borny ZN, Baki AKM (2021) A modified method of filtering for FBMC based 5G communications on minimizing doppler shift. In: 6th International conference for convergence in technology (I2CT), pp 1–4
7. Yadav A, Sethi D, Kumar S, Gurjar SL (2015) L and U slot loaded UWB microstrip antenna: C-band/WLAN notched. In: IEEE International conference on computational intelligence & communication technology, pp 380–384
8. Butenko VV, Devyatkin EE, Ivankovich MV, Lobov EM, Varlamov VO (2021) Modeling the physical layer of advanced radio access systems with increased spectral efficiency based on FBMC technology. In: Systems of signal synchronization, generating and processing in telecommunications (SYNCHROINFO), pp 1–5
9. Jang S, Na D, Choi K (2020) Comprehensive performance comparison between OFDM-based and FBMC-based uplink systems. In: International conference on information networking (ICOIN), pp 288–292
10. Choudhary G, Sethi D (2023) From conventional approach to machine learning and deep learning approach: an experimental and comprehensive review of image fusion techniques. Arch Comput Methods Eng 30(2):1267–1304
11. Kumar TA, Anjaneyulu L (2020) Channel estimation techniques for multicarrier OFDM 5G wireless communication systems. In: IEEE 10th international conference on system engineering and technology (ICSET), pp 98–101

12. Park B, Ryu HG (2019) Design and performance evaluation of SW (single extension windowing)-based CP-OFDM system for the sharp OOB spectrum. In: IEEE 2nd 5G world forum (5GWF), pp 508–511
13. Wei S, Li H, Han G, Zhang W, Luo X (2019) PAPR reduction of SLM-OFDM using Helmert sequence without side information. In: 14th IEEE conference on industrial electronics and applications (ICIEA), pp 533–536
14. Huang T, Hong W, Cai T, Lin H (2020) RLS-DPD algorithm for hybrid precoding architecture in MIMO-OFDM systems. In: IEEE 3rd international conference on electronic information and communication technology (ICEICT), pp 20–24
15. Zeng S, Wu P, Xia M (2021) An improved partial transmit sequence scheme for PAPR reduction in FBMC/OQAM systems. In: International wireless communications and mobile computing (IWCMC), pp 1259–1264
16. Dhua S, Arjun R, Appaiah K, Gadre VM (2020) Low complexity FBMC with filtered OFDM for 5G wireless systems. In: International conference on signal processing and communications (SPCOM), pp 1–5
17. Sheikh JA, Mir ZI, Mufti MUA, Parah SA, Bhat GM (2020) A new filter bank multicarrier (FBMC) based cognitive radio for 5G networks using optimization techniques. Wireless Pers Commun 112:1265–1280
18. Yadav A, Sethi D, Khanna RK (2016) Slot loaded UWB antenna: dual band notched characteristics. AEU-Int J Electron Commun 70(3):331–335
19. Chang YK, Ueng FB, Shen YS, Liao CH (2019) Joint channel estimation and turbo equalisation for MIMO-OFDM-IM systems. Int J Electron 106(5):721–740
20. Ojaroudi Parchin N, Jahanbakhsh Basherlou H, Al-Yasir YI, Abd-Alhameed RA, Abdulkhaleq AM, Noras JM (2019) Recent developments of reconfigurable antennas for current and future wireless communication systems. Electronics 8(2):128
21. Wen F, Wymeersch H, Peng B, Tay WP, So HC, Yang D (2019) A survey on 5G massive MIMO localization. Digital Sign Process 94:21–28
22. Li Y, Luo Y, Yang G (2019) High-isolation 3.5 GHz eight-antenna MIMO array using balanced open-slot antenna element for 5G smartphones. IEEE Trans Antennas Propag 67(6):3820–3830
23. Hesham H, Ismail T (2022) Hybrid NOMA-based ACO-FBMC/OQAM for next-generation indoor optical wireless communications using LiFi technology. Opt Quant Electron 54(3):201
24. Yapici Y, Guvenc I (2019) Non-orthogonal multiple access for mobile VLC networks with random receiver orientation. In: IEEE global communications conference (GLOBECOM), pp 1–6
25. Choudhary G, Sethi D (2023) Mathematical modeling and simulation of multi-focus image fusion techniques using the effect of image enhancement criteria: a systematic review and performance evaluation. Artif Intell Rev 56(11):13787–13839
26. Kumar A, Chakravarty S, Suganya S, Sharma H, Pareek R, Masud M, Aljahdali S (2022) Intelligent conventional and proposed hybrid 5G detection techniques. Alex Eng J 61(12):10485–10494
27. Mohammed A, Ismail T, Nassar A, Mostafa H (2021) A novel companding technique to reduce high peak to average power ratio in OFDM systems. IEEE Access 9:35217–35228
28. Jeya R, Amutha B (2019) Optimized semiblind sparse channel estimation algorithm for MU-MIMO OFDM system. Comput Commun 146:103–109
29. Chen-Hu K, Armada AG (2019) Low-complexity computation of zero-forcing equalizers for massive MIMO-OFDM. In: IEEE 89th vehicular technology conference (VTC2019-Spring), pp 1–5

Integration of Convolutional Neural Networks for Autonomous Guidance in Automated Guided Vehicles

Mukul Kumar, Sourabh Anand, Pushpendra S. Bharti, Manoj Kumar Satyarthi, Parveen Kumar, and Ajay Kumar

Abstract Integrating Convolutional Neural Networks with Autonomous Guided Vehicles offers an innovative method to improve industrial logistics. Conventional AGV navigation techniques have difficulties in adjusting to changing settings, necessitating the development of novel solutions. By utilizing CoppeliaSim, a robotics simulation platform, along with Python modules like OpenCV, NumPy, and Keras, it is possible to train and evaluate AGVs in simulated settings before their actual implementation in the real world. A recent study conducted training on a CNN model that consisted of a single hidden layer. The model included a total of 3,25,84,963 trainable parameters. The model attained an accuracy of 86.54% after 100 training epochs at a learning rate of 0.1. The precision, recall, and F1-score measures were captivating, indicating the model's ability to manoeuvre around obstacles with great effectiveness. The integration of CNN-driven navigation systems into AGVs has the potential to improve operating efficiency and flexibility in dynamic contexts.

Keywords Convolutional neural networks · Automated guided vehicles · Industrial automation machine learning · Real-time navigation · Dynamic environment · Autonomous guidance

M. Kumar · S. Anand · P. S. Bharti · M. K. Satyarthi
University School of Information, Communication and Technology, Guru Gobind Singh Indraprastha University, Sec 16C, New Delhi 110078, Dwarka, India

P. Kumar
Department of Mechanical Engineering, Rawal Institute of Engineering and Technology, Faridabad, Haryana, India

A. Kumar (✉)
Department of Mechanical Engineering, School of Engineering and Technology, JECRC University, Jaipur, Rajasthan, India
e-mail: ajay.kumar30886@gmail.com

A. Kumar et al. (eds.), *Advances in Mechatronics Systems*, Emerging Trends in Mechatronics, https://doi.org/10.1007/978-981-97-8472-1_7

1 Introduction

Automated Guided Vehicles (AGVs) are now essential assets to modern industrial logistics, providing a wide range of capabilities that optimize material handling and transportation operations [1, 2]. AGVs are autonomous vehicles designed for use in warehouse and factory settings. They are outfitted with sensors, navigation systems, and control mechanisms that allow them to function without a driver [3]. These vehicles are intricately engineered to negotiate complex layouts, deliver commodities, and execute diverse jobs with accuracy and efficiency. One of the main functions of AGVs is to convey materials [4, 5]. They function as very effective conveyors, transporting items, materials, and components between various places within a plant [6]. This includes the transportation of raw materials to manufacturing lines, the delivery of finished goods to storage places, and the movement of commodities to shipping terminals for distribution [7]. AGVs are designed with the capability to efficiently manage and transport loads of different dimensions and masses, thereby offering flexible and adaptable solutions for a wide range of logistical requirements [8–10]. AGVs play a crucial role in facilitating assembly line operations [11]. Their role involves transporting parts and components to production lines, facilitating smooth manufacturing operations, and reducing production downtime. AGVs may be configured to adhere to exact paths, ensuring the timely delivery of materials for assembly. This capability enhances workflow efficiency and minimizes periods of immobility. Furthermore, AGVs are capable of performing various jobs such as replenishing materials, removing garbage, and managing logistics at the production line, hence enhancing overall productivity and cost-efficiency. Advanced navigation systems have become essential in AGVs due to the growing complexity and dynamism of industrial surroundings. Conventional AGV navigation techniques, which depend on previously established paths or fixed maps, are unable to adapt to the changing requirements of modern industrial logistics. These approaches frequently lead to inefficiencies, delays, and safety risks, especially in settings with regular alterations in layout, production flow, and inventory positioning.

Machine learning (ML) techniques and Convolutional Neural Networks (CNNs) have become exceptionally efficient instruments for sensing and decision-making in diverse fields such as computer vision, robotics, and autonomous systems [12–16]. CNNs are a specialized category of advanced machine learning models that are particularly engineered to handle and interpret visual input. As a result, they are very effective for tasks like identifying objects, classifying images, and understanding scenes. CNNs have a hierarchical architecture that is inspired by the structure of the visual cortex in the human brain. This architecture allows them to extract important information from raw sensory inputs and learn intricate patterns and correlations within the data. In the domain of AGVs, CNNs provide revolutionary skills for perceiving and making decisions, allowing them to accurately understand and react to their environment [17]. CNNs may utilize sensor data, including camera pictures, LiDAR scans, and proximity sensor readings, to detect obstacles, navigate through complex settings, and make well-informed choices on route planning and trajectory control [18]. Furthermore, Convolutional Neural Networks CNNs have the capacity to adjust flexibly to variations in the surroundings, acquire knowledge via practice, and enhance their efficiency gradually, thus strengthening the resilience and dependability of AGV navigation systems [19, 20]. Conventional AGV navigation methods, which frequently depend on pre-established routes, magnetic strips, or floor markers that are integrated into the surface have many inherent constraints. These solutions are not sufficiently adaptable to accommodate dynamic changes in the environment, such as unforeseen impediments, changing layouts, or fluctuating traffic patterns. As a result, AGVs that rely on these conventional systems often experience frequent interruptions, which need manual intervention to redirect or modify their navigation trajectories. This not only diminishes the effectiveness of material handling activities but also amplifies the likelihood of accidents and operational downtime. Modern industrial environments are distinguished by their dynamic and frequently unforeseeable nature. Layouts may undergo regular changes in order to adapt to new operations, equipment, or storage needs. Consequently, AGVs need to possess the capability to promptly adjust to these modifications in order to maintain operational effectiveness. In order to achieve these flexibility requirements, it is crucial to have advanced navigation systems that are supported by technologies like CNNs. By implementing advanced algorithms, these technologies empower AGVs to autonomously analyze and react to their environment, guaranteeing optimal navigation performance in intricate and unpredictable scenarios.

In order to get an additional overview of the progress and influence of these developments, a concise review of the relevant literature is provided in the subsequent sections. This review examines the integration of machine learning, the rise of CNNs, and the most recent developments in AGV navigation systems.

1.1 Literature Review

Li et al. [21] developed a novel method to improve the effectiveness of multi-AGV systems in the manufacturing industry. Their work focuses on the fundamental problem of optimizing AGV operations while simultaneously avoiding deadlocks. Their work suggests a hybrid approach that combines a conventional AGV scheduling algorithm with an artificial neural network (ANN). This ANN component is specifically developed to forecast upcoming activities and proactively allocate AGVs to the initial locations of these jobs. The suggested technique enhances system efficiency by integrating the strengths of classical scheduling algorithms for deadlock prevention with the predictive capabilities of deep learning. The simulation findings indicate that this integrated approach improves efficiency by 20% to 30% compared to traditional approaches, resulting in a significant enhancement in the performance of AGV systems.

Dias et al. [22] addressed the challenge of developing autonomous robots capable of obstacle avoidance. Their study applied fuzzy logic to enhance the decision-making processes of autonomous robots in dynamic environments. The key input parameters included distance to obstacles and the robot's speed and direction. By implementing a fuzzy logic controller, the authors successfully developed an autonomous robot that can effectively navigate and avoid obstacles. Their experimental results demonstrated that the fuzzy logic-based approach significantly improves the robot's ability to make real-time navigational decisions, ensuring smoother and more efficient path planning compared to traditional methods. This research contributes to the advancement of autonomous robotic systems by providing a robust method for obstacle deviation.

Ngo and Tran [23] proposed a control strategy based on fuzzy logic for AGVs in order to improve the accuracy of path-following under different load scenarios. Their study highlighted the effectiveness of the fuzzy logic technique in adapting to load variations, hence enhancing the tracking performance of the AGV and assuring its dependable functioning in a wide range of industrial settings.

Zacharia and Xidias [24] established a genetic algorithm that employs fuzzy logic to determine the optimal routes and motion plans for AGVs in flexible manufacturing environments. Their methodology combines fuzzy logic for handling uncertainties with a genetic algorithm for identifying ideal routes, so greatly improving the efficiency and flexibility of AGV routing in dynamic contexts.

Kafiev et al. [25] presented an AGV fuzzy logic-based control system. Increasing AGV navigation and decision-making skills is the main goal of their work. The control system in different industrial environments efficiently controls path planning and obstacle avoidance by using fuzzy logic, which improves overall operational efficiency and dependability.

Using a four-layer neural network, Yuan et al. [26] studied and developed an intelligent vehicle path planning system. Their work intended to improve intelligent vehicle path planning accuracy and efficiency. The four-layer neural network enhances the navigation ability of the vehicle in dynamic and unpredictable situations by efficiently processing complicated environmental input to provide ideal routes.

Based on a Radial Basis Function (RBF) neural network, Wong and Yu [27] developed an anti-disturbance vSLAM (visual Simultaneous Localization and Mapping) navigation method for mobile robots. Their method seeks to improve the robot's stability and precision of navigation in the presence of environmental disturbances. The approach enhances the robot's capacity to sustain dependable localization and mapping by using the RBF neural network, hence improving overall navigation performance in complicated settings.

Using an ANN working together with the A* algorithm, Huu et al. [28] created a dynamic path-planning system. Their method significantly enhances the efficiency and flexibility of routing in dynamic settings by integrating ANN for real-time modifications and optimization with A* for initial path planning.

Teso-Fz-Betoño et al. [29] created an interior navigation system for an autonomous mobile robot. Their method makes use of semantic segmentation techniques to improve the robot's comprehension and navigation in indoor settings, hence raising the precision and effectiveness of autonomous navigation.

Applying deep neural networks, Ren et al. [30] developed an AGV real-time optimum navigation system. They use real-time obstacle detection and path planning algorithms among other methods for managing both static and dynamic impediments. The AGV's capacity to move quickly and securely in complicated settings is much improved by this integration.

For autonomous vehicles navigating unfamiliar terrain, Biswas et al. [31] introduced a particle swarm optimization (PSO) based path planning technique. With this method, autonomous systems should be able to navigate more effectively and safely across complicated and foreign settings.

Yu and Yan [32] approach enhances the particle swarm optimization (PSO) algorithm for AGV path planning. Probably, this improvement is adjusting the PSO algorithm's settings and methods to better fit the particular needs and limitations of AGV navigation jobs. By modifying the PSO algorithm, the suggested approach seeks to enhance the resilience, convergence speed, and quality of path-planning solutions for AGVs.

Song [33] carried out a global path-planning technique for Automatic Guided Vehicles that takes speed control and multi-objective optimization. The work intends to use speed control techniques and concurrently take into account several goals in order to improve AGV routes. Through the integration of these elements, the suggested approach improves the efficacy and efficiency of AGV navigation in complicated settings.

Al-Jarrah et al. [34] used a bacterial algorithm with fuzzy image processing to present a new detection method for vision systems. The method generates 2D maps for navigation and precisely recognizes the robot and obstacles in real-time robot navigation tests. Time-based trajectories are produced for best path planning by a modified genetic algorithm (GA). Different interior conditions confirm the effectiveness of the algorithm in computation and detection.

In a recent study conducted by Jiang et al. [35], a novel adaptive genetic algorithm was proposed for the purpose of mobile robot path planning. This algorithm bears resemblance to the well-known ordered clustered Traveling Salesman Problem (TSP). The proposed methodology incorporates a priority matrix that is derived from station priorities, along with improvements to the initial population generation process in order to mitigate the occurrence of unfeasible solutions. Furthermore, the utilization of adaptive operators and simulated annealing strategies is implemented in order to circumvent local optima and expedite the process of convergence. The empirical findings substantiate the efficacy of the algorithm in circumventing local optima and attaining expedited convergence.

The observed research gap, as derived from the literature stated above, pertains to the constraints of conventional AGV navigation techniques in their ability to effectively adjust to dynamic and evolving environments. The integration of CNNs with AGVs holds significant promise in improving industrial logistics. However, current navigation techniques encounter challenges when it comes to effectively manoeuvring AGVs in diverse environments characterized by changing conditions and obstacles. The aforementioned gap underscores the necessity for strong and flexible navigation solutions that are specifically designed for AGVs, especially in dynamic industrial environments. Hence, it is imperative to conduct additional research in order to formulate novel methodologies that effectively address these constraints and enhance the effectiveness and adaptability of AGV navigation systems. In order to address the existing research gap pertaining to the limitations of conventional AGV navigation techniques, the proposed approach places emphasis on the development of resilient and flexible navigation solutions specifically designed for AGVs, with a particular emphasis on dynamic industrial environments as visualized through Fig. 1a–d. By harnessing advanced tools such as CoppeliaSim and Python modules like OpenCV, NumPy, and Keras, this study aims to create and assess a CNN-driven navigation system for AGVs. After conducting a series of training epochs, the CNN model, which consists of a grand total of 3,25,84,963 trainable parameters, has demonstrated remarkable performance. After 100 training epochs at a learning rate of 0.1, the model has achieved an impressive accuracy of 86.54%. This achievement highlights the model's ability to effectively perceive and navigate intricate environments, while also adapting to dynamic changes and obstacles. The evaluation of the potential consequences of incorporating CNN-driven navigation systems into AGVs seeks to offer valuable insights into enhancing operational efficiency and adaptability in practical industrial settings. This endeavour seeks to bridge the existing research gap and propel the advancement of AGV navigation performance.

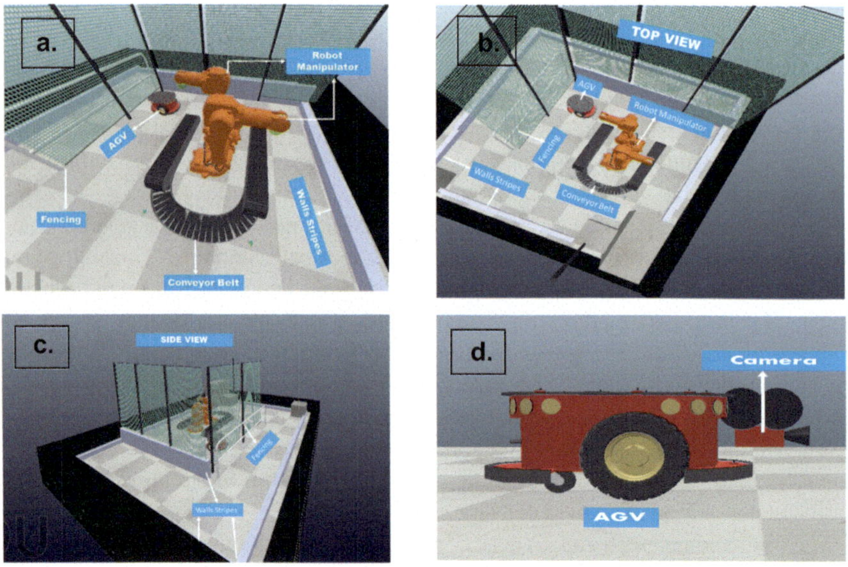

Fig. 1 Industrial work cell environment where the AGV operates, showing **a** isometric view, **b** top view, **c** side view. **d** AGV with perspective Cameral

2 Development of CNN Model

The CNN model, which was developed in the present study, was purposefully designed for the purpose of image classification jobs, employing a customized dataset which has been shown through a flow diagram (Fig. 3). The images were acquired using an AGV that was fitted with a vision sensor. The camera recorded photos from several viewpoints (left, right, front) (shown in Fig. 2) of the AGV's surroundings inside an industrial work cell. In order to provide a thorough comprehension of the arrangement, the precise characteristics of the camera used are outlined in Table 1. The dataset was comprised of grayscale images, which were resized to 50 × 50 pixels in order to maintain consistency and compatibility with the input layer of the model.

The grayscale images were first imported and subsequently resized to dimensions of 50 × 50 pixels. The pixel values were then normalized by dividing them by 255. The implementation of this normalization step was carried out to ensure that the pixel values were appropriately scaled within the range of 0 to 1, a critical factor in facilitating the optimal training of the neural network. The images underwent a

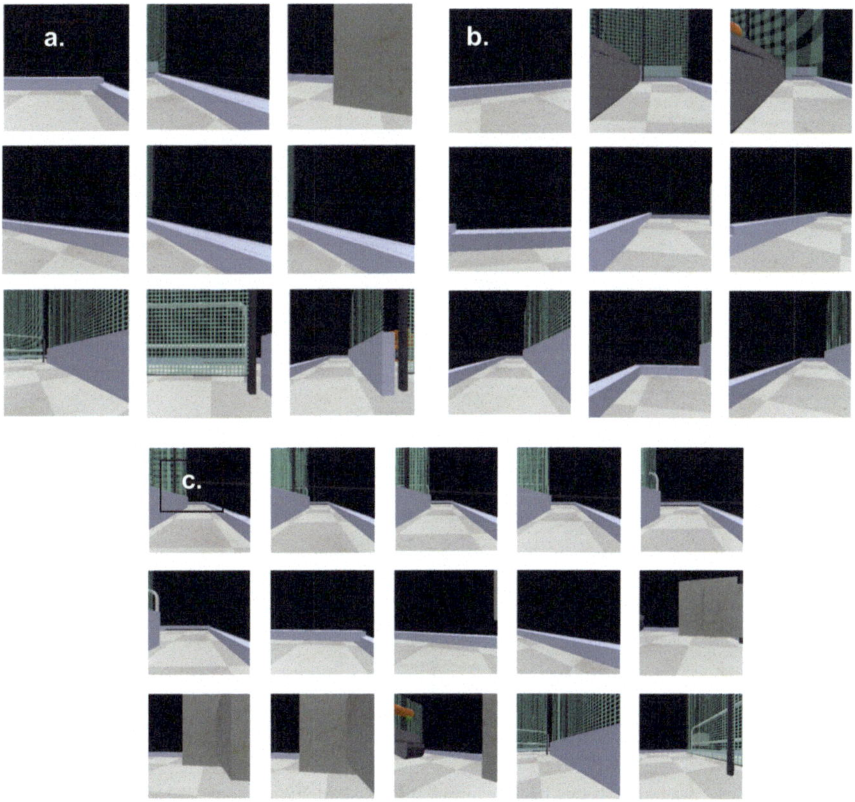

Fig. 2 Example images from the dataset: **a** Left view, **b** Right view, **c** Front view

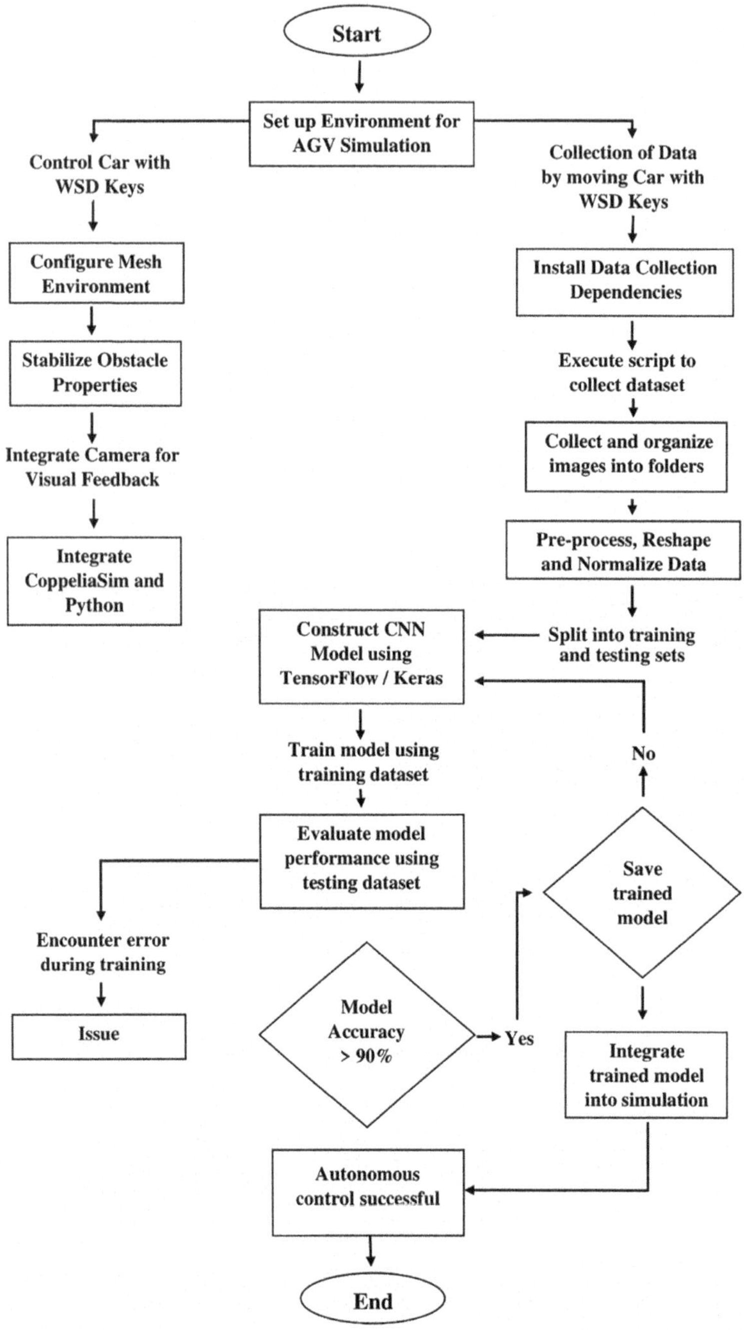

Fig. 3 Workflow diagram of the CNN training process, including data preprocessing, model training, and evaluation

Table 1 Camera specifications

Parameters	Value/type
Camera perspective	Perspective
Near/far clipping plane (m)	0.01/5.00
Perspective. Angle [degree]/Ortho. size [m]	78
Resolution X/Y	256/256
Sensor size [m]	0.01

process of reshaping to incorporate a solitary channel, thereby yielding a data shape denoted as $(n, 50, 50, 1)$, where n signifies the quantity of images contained within the dataset. The labels were transformed into a categorical format through the application of one-hot encoding, thereby enabling the process of multi-class classification.

The CNN architecture (as shown in Fig. 4) includes a pair of convolutional layers, a flatten layer, a dense hidden layer incorporating dropout regularization, and an output dense layer employing SoftMax activation [36]. In order to extract low-level features from the input images, the initial convolutional layer employed 256 filters with a size of 3×3. The convolution operation is mathematically represented as [37, 38]:

$$\mathrm{Z} = \mathrm{X}^*\mathrm{W} + \mathrm{b} \quad (1)$$

The variable X represents the input tensor, while W denotes the filters and b signifies the bias term. Subsequent to the convolution operation, the model underwent a Rectified Linear Unit (ReLU) activation function, thereby introducing non-linearity. The Rectified Linear Unit (ReLU) function is mathematically defined as:

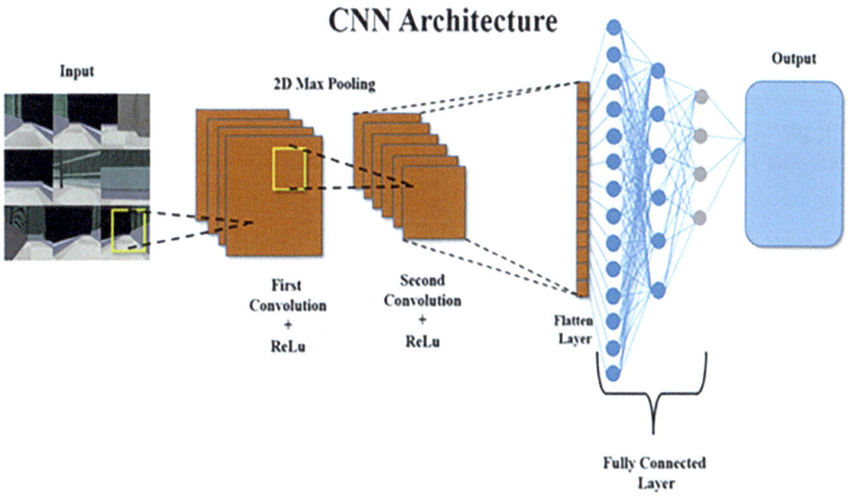

Fig. 4 CNN model development

$$A = \mathrm{RelU}(Z) = \max(0, Z) \tag{2}$$

The application of a MaxPooling2D layer, with a pool size of 2 × 2, was subsequently employed to decrease the spatial dimensions of the feature maps by 50%. This reduction in dimension aimed to decrease computational measuring 3 × 3 and retrieved more intricate patterns from the feature maps produced by the first convolutional layer.

The second convolutional layer, consisting of 128 filters. Subsequently, there was an additional implementation of the ReLU activation function and a MaxPooling2D layer, resulting in further reduction of the spatial dimensions of the feature maps. The final dense layer relied on the SoftMax activation function to generate a probability distribution across the different classes, thereby facilitating multi-class classification. The SoftMax function is a mathematical function commonly used in machine learning and statistics. It is defined as:

$$\mathrm{SoftMax}(Z_i) = \frac{e^{Z_i}}{\sum_j e^{Z_j}} \tag{3}$$

The model was trained using the Adam optimizer, which was selected for its adaptive learning rate capabilities. The learning rate (η) was set to 0.1. The learning rate is a crucial hyperparameter that determines the size of the updates made to the model parameters during the training process. The parameter update rule is as follows:

$$\theta_{t+1} = \theta_t - \eta \nabla_\theta L(\theta) \tag{4}$$

where $L(\theta)$ represents the loss function, the training process consisted of 100 epochs, during which the model weights were updated iteratively based on the training data. The study aimed to minimize the loss function by optimizing the model parameters using the backpropagation technique. Throughout the training process, we evaluated multiple performance metrics including accuracy, precision, recall, and F1-score to assess the learning progress and effectiveness of the model.

3 Result and Discussion

The incorporation of a CNN into AGVs for self-guided navigation showed great potential, as shown through multiple performance measures. The CNN architecture consisted of a single hidden layer and an aggregate of 32,584,963 trainable parameters. The model was tuned over 100 training epochs, using a learning rate of 0.1. Throughout the training phase, the model's performance was consistently examined and documented to enable efficient learning and generalization.

3.1 Training Process and Performance Metrics

The training phase included optimizing the model weights to reduce the discrepancy between the expected and actual results. Throughout the 100 epochs, the model's accuracy demonstrated a steady and gradual improvement, suggesting successful learning. Simultaneously, the loss function, which measures the model's prediction mistakes, continually declined. The model attained a final accuracy of 86.54%, demonstrating its ability to accurately categorize the majority of pictures (Table 2). The model's efficacy in eliminating prediction errors is shown by the ultimate loss value of 1.1211 (Fig. 5). The model's precision, defined as the ratio of genuine positive predictions to all positive predictions, was 0.6704. The recall, which measures the model's capacity to correctly identify positive events, was 0.6217. The F1-score, which is calculated as the harmonic mean of accuracy and recall, was 0.6404. These measures together highlight the model's ability to effectively manage both false positives and false negatives, demonstrating a well-balanced performance.

Table 2 Performance metrics of the CNN model

Metric	Value
Accuracy	0.8654
Loss	1.1211
Precision	0.67054
Recall	0.6217
F1-score	0.6404

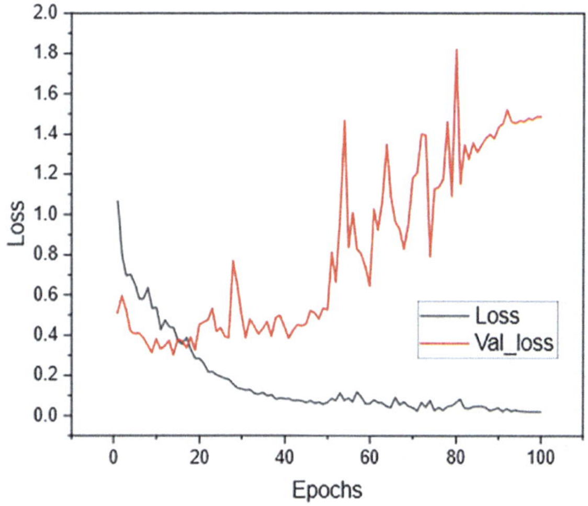

Fig. 5 Loss versus epochs

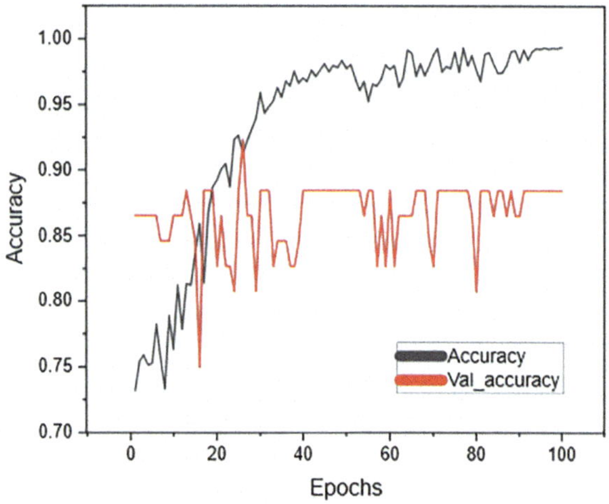

Fig. 6 Accuracy versus epochs

3.2 Accuracy and Loss Curves

The graph in Fig. 6 illustrates a consistent rise in accuracy throughout the early epochs, followed by a stabilization towards the conclusion of the training period. This suggests that the model was successfully acquiring knowledge and applying it to new data. The loss versus epochs graph (Fig. 5) demonstrates a steady decline in the loss function, providing more evidence of the model's successful learning process.

3.3 Application in AGV Navigation

The CNN model that has undergone training was used to improve the self-guided movement of an AGV in an industrial work cell. The navigation of the AGV was dependent on the model's capacity to categorize photos taken from different angles, enabling it to travel and carry out tasks inside the work cell. The model's strong performance, as seen by the training measures, resulted in its successful navigation skills in real-world scenarios. A video of the AGV successfully traversing the industrial work cell using the trained CNN model for image classification is available for viewing via the provided link. The video demonstrates the actual use and efficiency of the model in a dynamic industrial setting, emphasizing its capability for immediate operational implementation [39–43].

4 Conclusion

The integration of a CNN for the self-directed navigation of AGVs in an industrial work cell has shown considerable promise and efficiency. The CNN model, consisting of 1 hidden layer and 32,584,963 trainable parameters, demonstrated strong performance metrics. These include an accuracy of 86.54%, a loss of 1.1211, a precision of 0.6704, a recall of 0.6217, and an F1-score of 0.6404. The training procedure spanned 100 epochs, with a learning rate of 0.1, which resulted in successful learning and generalization. This is supported by the persistent rise in accuracy and continual decrease in loss. The use of this model greatly improved the AGV's navigation skills, allowing it to accurately categorize photos from different angles and perform duties inside the industrial work cell. The extensive performance measurements and visualizations highlight the model's capability for immediate practical implementation, making it a vital instrument for enhancing the effectiveness and precision of AGV navigation in ever-changing industrial settings. The effective implementation of the CNN model in this study emphasizes the significance of sophisticated machine learning methods in industrial automation and establishes a basis for future research and development in this domain. Continued enhancements and modifications of this model have the potential to result in more advanced and effective AGV systems, therefore contributing to the progress of intelligent manufacturing and automated logistics.

References

1. Kubasakova I et al (2024) Implementation of automated guided vehicles for the automation of selected processes and elimination of collisions between handling equipment and humans in the warehouse 24(3):1029
2. Ajay M, Kant R (2020) Incremental sheet forming technologies: principles, merits, limitations, and applications, 1st edn. CRC Press. https://doi.org/10.1201/9780429298905
3. Singh H, AlMangour B (eds) (2023) Handbook of smart manufacturing: forecasting the future of industry 4.0, 1st edn. CRC Press. https://doi.org/10.1201/9781003333760
4. Torchio F (2023) Survey on automated systems for smart warehouses. Politecnico di Torino
5. Kumar A, Liu Y, Kumar R (eds) (2024) Handbook of intelligent and sustainable manufacturing: tools, principles, and strategies, 1st edn. CRC Press. https://doi.org/10.1201/9781003405870
6. Aized T (2010) Materials handling in flexible manufacturing systems. Future Manuf Syst:121–136
7. Rooks B (2001) AGVs find their way to greater flexibility. Assemb Autom 21(1):38–43
8. Kumar A, Rajak DK, Kumar P, Rathee S (eds) (2024) Smart electric and hybrid vehicles: design, modeling, and assessment by industry 4.0 approaches, 1st edn. CRC Press. https://doi.org/10.1201/9781003502470
9. Kumar A, Rajak DK, Kumar P, Rathee S (eds) (2024) Smart electric and hybrid vehicles: fundamentals, strategies and applications, 1st edn. CRC Press. https://doi.org/10.1201/9781003495574
10. Anand S, Satyarthi M (2024) Robotic manipulator, assisted fused filament fabrication process. In: AIP conference proceedings. AIP Publishing
11. Vlachos IP et al (2023) Lean manufacturing systems in the area of Industry 4.0: a lean automation plan of AGVs/IoT integration 34(4):345–358

12. Sony S et al (2021) A systematic review of convolutional neural network-based structural condition assessment techniques 226:111347
13. Rani S, Tripathi K, Kumar A (2023) Machine learning aided malware detection for secure and smart manufacturing: a comprehensive analysis of the state of the art. Int J Interact Des Manuf (IJIDeM). https://doi.org/10.1007/s12008-023-01578-0
14. Mohammadi M, Jamshidi S, Rezvanian A, Gheisari M, Kumar A (2024) Advanced fusion of MTM-LSTM and MLP models for time series forecasting: an application for forecasting the solar radiation. Meas Sensors 33:101179. https://doi.org/10.1016/j.measen.2024.101179
15. Kumar A, Shrivastava VK, Kumar P, Kumar A, Gulati V (2024c) Predictive and experimental analysis of forces in die-less forming using artificial intelligence techniques. Proc Instit Mech Eng Part E J Process Mech Eng. https://doi.org/10.1177/09544089241235473
16. Burande DV, Kalita K, Gupta R, Kumar A, Chohan JS, Kumar D (2024) Machine learning metamodels for thermo-mechanical analysis of friction stir welding. Int J Interact Des Manuf (IJIDeM). https://doi.org/10.1007/s12008-024-01871-6
17. Lynch Z, Laursen B (2009) The neuro revolution: how brain science is changing our world. St. Martin's Press
18. Melotti G, Asvadi A, Premebida C (2018) CNN-LIDAR pedestrian classification: combining range and reflectance data. In: IEEE International conference on vehicular electronics and safety (ICVES). IEEE
19. Zhang J et al (2023) Automated guided vehicles and autonomous mobile robots for recognition and tracking in civil engineering 146:104699
20. Han Z (2023) Multimodal intelligent logistics robot combining 3D CNN, LSTM, and visual SLAM for path planning and control. Front Neurobot 17:1285673
21. Li D et al (2019) Artificial intelligence empowered multi-AGVs in manufacturing systems
22. Dias LA et al (2018) Application of the fuzzy logic for the development of automnomous robot with obstacles deviation 16(2):823–833
23. Ngo HQT, Tran AS (2018) Using fuzzy logic scheme for automated guided vehicle to track following path under various load. In: 4th International conference on green technology and sustainable development (GTSD). IEEE
24. Zacharia PT, Xidias EK (2020) AGV routing and motion planning in a flexible manufacturing system using a fuzzy-based genetic algorithm. Int J Adv Manuf Technol 109(7):1801–1813
25. Kafiev I, Romanov P, Romanova I (2020) Fuzzy logic based control system for automated guided vehicle. In: International multi-conference on industrial engineering and modern Technologies (FarEastCon). IEEE
26. Yuan H et al (2019) Research and implementation of intelligent vehicle path planning based on four-layer neural network. In: Chinese automation congress (CAC)
27. Wong SF, Yu Z (2019) The mobile robot anti-disturbance vSLAM navigation algorithm based on RBF neural network. Procedia Manuf 38:400–407
28. Huu TT, Duc DN, Nananukul N (2020) Machine learning model for a dynamic path planning problem. J Phys: Conf Ser (IOP Publishing)
29. Teso-Fz-Betoño D et al (2020) Semantic segmentation to develop an indoor navigation system for an autonomous mobile robot 8(5):855
30. Ren Z et al (2021) Deep neural networks-based real-time optimal navigation for an automatic guided vehicle with static and dynamic obstacles. Neurocomputing 443:329–344
31. Biswas S, Anavatti SG, Garratt MA (2019) A particle swarm optimization based path planning method for autonomous systems in unknown terrain. In: IEEE international conference on industry 4.0, artificial intelligence, and communications technology (IAICT). IEEE
32. Yu X, Yan J (2019) AGV path planning based on optimized particle swarm optimization. In: 3rd International conference on electronic information technology and computer engineering (EITCE). IEEE
33. Song JJS (2021) Automatic guided vehicle global path planning considering multi-objective optimization and speed control. Materials 33
34. Al-Jarrah R, Al-Jarrah M, Roth H (2018) A novel edge detection algorithm for mobile robot path planning. J Robot

35. Jiang J et al (2020) An improved adaptive genetic algorithm for mobile robot path planning analogous to the ordered clustered TSP. In: IEEE congress on evolutionary computation (CEC). IEEE
36. Alzubaidi L et al (2021) Review of deep learning: concepts, CNN architectures, challenges, applications, future directions 8:1–74
37. Kuo CCJ (2016) Understanding convolutional neural networks with a mathematical model. J Vis Commun Image Represent 41:406–413
38. Wiatowski T, Bölcskei H (2017) A mathematical theory of deep convolutional neural networks for feature extraction. IEEE Trans Inf Theory 64(3):1845–1866
39. Engineer R (2024) CoppeliaSim simulation—AGV CNN model. https://www.youtube.com/watch?v=0DuRYmGo6CY&list=LL&index=2. Accessed 2 May 2024
40. Chauhan S, Cheruku R, Edla DR, Kampa L, Nayak SR, Giri J, Mallik S, Aluvala S, Boddu V, Qin H (2024) BT-CNN: a balanced binary tree architecture for classification of brain tumour using MRI imaging. Front Physiol 15. https://doi.org/10.3389/fphys.2024.1349111
41. Mahakalkar SG, Giri JP, Sunheriya N, Lande P, Chdge R, Mahatme C, Sathish T (2024) Propulsion effectiveness using propeller selection, ducted propellers and tilting mechanism in a quadrotor drone. Lecture notes in mechanical engineering, pp 399–410. https://doi.org/10.1007/978-981-97-3173-2_26
42. Khan SWA, Dapke A, Giri J, Sunheriya N (2023) A comparative study of a plane NACA2412 airfoil with a NACA2412 airfoil attached with gurney flap using CFD. CRC Press, pp 1392–1402. https://doi.org/10.1201/9781003370628-74
43. Qureshi MJ, Kutty AV, Giri J., Dapke A, Giri P, Chadge RB, Sunheriya N (2023) Public acceptance of drones: an approach. CRC Press, pp 493–503. https://doi.org/10.1201/9781003358596-55

Survey on Drones and IOT Devices for Futuristic Smart City

Satnam Kaur and Nancy Arya

Abstract Intelligent gadgets and technology that improve quality of life, advance sustainability, and save lives define smart cities. Drones and the Internet of Things (IoT) are examples of advanced collaboration technologies that are essential for reaching these goals. For Internet of Things (IoT) applications, high-quality service must be provided drone use is an effective strategy IoT technologies increase smart city intelligence by improving connection power-saving and quality of services (QoS). This study aims to demonstrate how cooperative drone technology and the Internet of Things (IoT) could improve the intelligence of smart cities. It works on a variety of issues, including data gathering, privacy and security, public safety, emergency preparedness, energy use, and lifestyle in smart cities. Drones and IoT in smart cities are mostly used for real-time applications. Indeed, drones have a lot of potential in both the public and civic spheres. They offer a variety of uses where human interaction could be dangerous or unfeasible. Future wireless networks can use drones to provide flexible wireless access, enabling bigger coverage areas and faster transmission rates.

Keywords Quality of services (QoS) · Internet of Things (IoT) · IoT gateway · Artificial intelligence (AI) · Internet of Drones (IoD)

1 Introduction

The supply of services including energy efficiency, air quality, and traffic surveillance, healthcare, safety for society, and disaster recovery are all intimately tied to smart cities. Artificial intelligence (AI), robots, and the Internet of Things (IoT) all play critical roles in raising cities' intelligence in order to accomplish these goals. The

S. Kaur (✉) · N. Arya
Department of CSE, SGT University, Gurgaon, Haryana, India
e-mail: Satnam_fet@sgtuniversity.org

N. Arya
e-mail: nancy_feat@sgtuniversity.org

A. Kumar et al. (eds.), *Advances in Mechatronics Systems*, Emerging Trends in Mechatronics, https://doi.org/10.1007/978-981-97-8472-1_8

infrastructure of smart cities is strengthened by these advances, which also improve environmental aspects including energy efficiency, air quality, and traffic monitoring as well as quality of life, healthcare, public safety, and disaster management and facilitate the delivery of smart services, ultimately enhancing the standard of living for residents.

The building of effective infrastructure that can deliver cost-effective services is a primary priority in the creation of smart cities. In order to create a pervasive and linked environment, it is important to make these services and amenities available throughout the entire city. Smart cities may maximize drone service delivery, cut operational costs, and improve resource efficiency by utilizing cutting-edge technologies.

Cities can manage utilities, transportation systems, healthcare services, emergency response systems, and environmental monitoring better by integrating drones with IOT. These technologies make it possible to make data-driven decisions and perform predictive analytics, automation, and real-time monitoring, which results in better service delivery and more effective resource management.

Smart cities aim to improve the general well-being of their citizens while achieving economic and environmental sustainability by embracing cutting-edge technologies and putting a strong emphasis on effective infrastructure.

An engineering paradigm known as the Internet of Things (IoT), permits intelligent connectivity between recognizable and well-known smart gadgets supporting the development of an all-pervasive and intelligent framework [1]. Physical things may now sense, analyze, and carry out tasks by establishing communication and information sharing among themselves thanks to the power of (IoT). This transformative technology enables these objects to transition from simple objects to intelligent entities by exploiting underlying improvements in internet protocols, sensor networks, devices that are embedded, ubiquitous and pervasive computing, and communication technologies [2].

Much research has been undertaken recently with the goal of enhancing smart city characteristics in a variety of areas and facets. Recent developments in drone technology have significantly helped the development of smart cities. In the modern world, without drone services, it is challenging to imagine a smart city [3].

Self-contained aerial robots known as drones have a number of practical uses in daily life. These programs are employed in a wide range of industries, including actions for gathering, monitoring, and reconnaissance of intelligence, transportation, agriculture, protection against disasters, safety, and security [3].

Figure 1 shows how drones and IoT are integrated cooperatively in a variety of smart city applications, such as gathering pollution data, forecasting air quality, and using data from IoT devices to increase energy efficiency. Drones and Internet of Things technology work together to save lives in the context of disaster management. It makes it possible to distribute necessities like food and blood in real time and to keep track of accidents and victims. This integration facilitates timely and efficient response efforts to mitigate the impact of disasters and ensure the well-being of affected individuals.

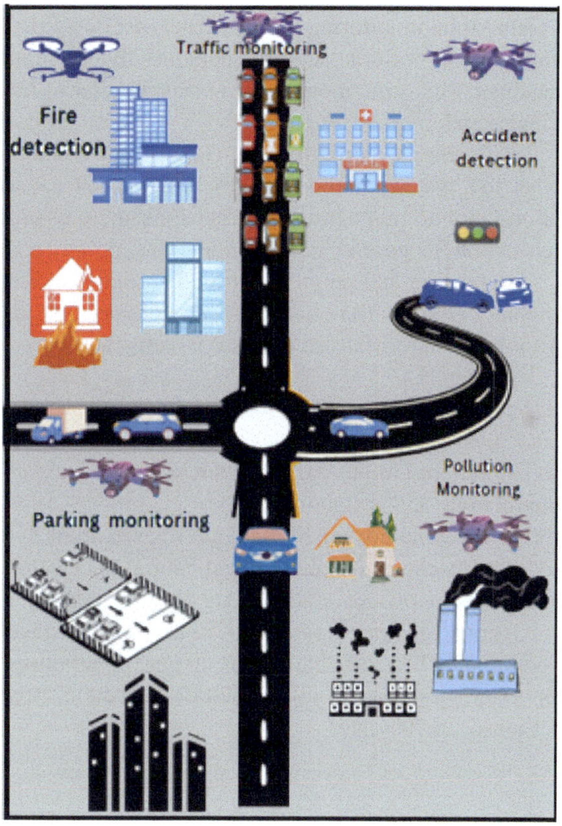

Fig. 1 Drone and Internet of Things collaboration to make smart cities smarter

2 Brief Overview of IoT and Its Integration with Drones

IoT stands for "Internet of Things" as it is commonly abbreviated, is a collection of networked physical objects like cars, appliances, and other vehicles that are also connected to the internet and have software built in. These gadgets have the ability to gather and exchange data, which allows them to speak and interact with people.

Unmanned aerial vehicles (UAVs), or drones, are flying objects that can be piloted remotely or automatically. They can carry out activities including aerial surveillance, monitoring, mapping, and distribution because of their array of sensors, cameras, and communication systems.

Drone-based systems have significantly advanced and gained new capabilities thanks to the Internet of Things (IoT) and drones. Drones can benefit from the IoT ecosystem in a number of ways.

Data Gathering: Using sensors, cameras, and other onboard devices, drones can gather a huge quantity of data about their surroundings. This data can be sent in real-time to centralized systems for processing and analysis by incorporating IoT.

Real-time monitoring: The Internet of Things (IoT) makes it feasible to continuously monitor drone activity, including flying parameters, battery life, and sensor readings. Real-time monitoring makes it easier to react quickly to shifting circumstances and improves safety.

Intelligent Decision-Making: Drones may communicate and exchange data with other IoT systems and devices by utilizing IoT capabilities. They can then use the information to make informed decisions about changing flight paths, working with sensors on the ground, or responding to certain situations.

Automation and remote control: Through centralized interfaces, IoT integration enables operators to remotely operate drones. Additionally, predetermined rules or data-driven algorithms can be used to automate processes, enabling drones to operate independently.

Scalability and cooperation: Connecting several drones to a single network is made possible by the Internet of Things (IoT), allowing for coordinated action and collaboration. Drones can collaborate, share data, and coordinate tasks to increase productivity and scalability.

Improved Safety and Security: IoT-based drone systems can include cutting-edge security mechanisms like authentication, encryption, and secure data transfer to guarantee the privacy and integrity of critical data collected by drones.

Overall, the integration of IoT with drones expands their capabilities, improves operational efficiency, and opens up new possibilities for applications in areas such as aerial surveillance, infrastructure inspection, precision agriculture, emergency response, and beyond.

3 Related Work

Numerous studies have examined how drones can contribute to enhancing the level of intelligence in smart cities across various applications.

Geetanjali et al. proposed trust establishment scheme is introduced for the ad hoc network of Unmanned Aerial Vehicles (UAVs). This scheme aims to enhance trust and reliability within the network, ensuring that communication remains secure and protected [4].

In order to achieve low latency and secure communication among drones within a smart city, it is crucial to establish a robust authentication process between the involved parties in each specific zone. In this paper, the author addressed the challenge of ensuring secure and low-latency authentication for drones by employing blockchain-based security mechanisms [5].

The evaluation of smart cities' intelligence takes into account multiple factors smart environment, smart lifestyle, smart transportation, smart tourism, and more. Another study by Abdelmaboud [6], explores the fundamental requirements of security, privacy, and communication in the context of the Internet of Drones (IoD). It introduces a comprehensive taxonomy of IoD, developed based on pertinent considerations, and provides an overview of commonly employed commercial case studies.

Additionally, the document discusses the most recent developments and suggested fixes designed especially for IoD environments. Finally, the authors analyze the difficulties now facing IoD and outline potential future research routes, shaping the trajectory of this field toward innovative solutions and further exploration [6]

The authors of Lai et al. [7] A reliable and efficient transportation system is a crucial element for any smart city to thrive. Intelligent Transportation Systems (ITS) Globally, smart cities have already adopted ITS technologies, and ITS development depends on the smooth integration of connected and autonomous cars. In several locations throughout the world, these two technologies are currently undergoing intensive field testing. Although the development of fully automated transportation systems depends on links and autonomy this paper examines the possible uses of UAVs in ITS, stressing the benefits and drawbacks that UAV-enabled ITS offers for the development of future smart cities [7].

Mohammad Aminul Hoque's study The utilization of advanced communication and computation technologies in IoT devices enhances the realization of the smart city vision. However, the deployment and management of numerous IoT devices and gateways in IoT-based smart city applications pose challenges in terms of complexity, time consumption, and expenses. To address these issues, a drone-based IoTaaS framework is proposed in this paper. This framework aims to optimize the deployment, cost, and usage of IoT devices. Additionally, It includes an economic analysis of IoT services using drones. The implementation of IoTaaS, maximizing resource utilization, and providing users with the greatest offers for satisfying their needs are all ways that The vendor of IoT infrastructure services can profitably rent out their equipment. IoTaaS has been used as a proof-of-concept in applications including monitoring of air pollution and intelligent agriculture, showing how it can save setup costs and maximize the use of IoT devices [8].

The authors Alsamhi et al. [9] study about transportation, agriculture, security and safety, disaster relief, environmental protection, service provision, energy savings, e-waste reduction, weather monitoring, healthcare, and other areas all benefit greatly from collaborative drones and IoT technologies. With an emphasis on their latest developments for boosting the intelligence of smart cities, this study presents a survey that investigates the prospective methods and uses of collaborative drones and IoT. The study offers a thorough overview of the most recent studies and ongoing initiatives in deploying cooperative drones and the Internet of Things to support real-time applications in smart cities. The report also identifies significant obstacles and provides research-based solutions to overcome them, seeking to increase the effectiveness of this collaboration [9].

Thomas Lagkas et al. [10] look at how the use of unmanned aerial vehicles (UAVs) is widespread. scenarios and use cases, along with how they are integrated with IoT sensors. The authors examine the possible support offered by cutting-edge technologies like 5G networks for drones as IoT devices. They review fleet management options employing aerial networking and examine the sensor needs for these applications.

There is much discussion of user and drone privacy and security concerns. The authors suggest a framework that makes it easier to integrate these technologies into

UAVs in order to allay these worries. Modern vision-based scene analysis technologies are incorporated into the framework, with a focus on strong security and privacy measures.

According to the suggested framework, UAV operations including registration, identification, rating, and the management of moving objects are controlled by a hybrid centralized-distributed architecture. The authors intend to assess the suggested framework in both lab and real-world contexts as part of their ongoing research. This assessment will assist in improving the framework to accommodate particular application domains based on contextual needs as mentioned in Table 1.

Kuan-Wen Chen introduces DroneTalk, a cutting-edge drone delivery system for mail services, harnessing the power of the Internet of Things (IoT). This system seamlessly integrates a GPS, an inertial measurement unit, and advanced visual technology, enabling a versatile autopilot mode capable of operating both indoors and outdoors. The system considers potential communication, processing, and control latencies-related drone collision problems. The suggested system's successful functionality has been demonstrated through careful simulations and field experiments [3].

Mohit Anguralaa This paper introduces the DBCMS mechanism, a defense system created to protect medical employees at risk of COVID-19 infection. The suggested strategy provides an effective way to improve patients with COVID-19's therapeutic process while protecting their security and wellbeing [12].

4 Market Opportunity

For equipment suppliers, Internet service providers, and software developers, the Internet of Things (IoT) presents a substantial business opportunity. By the end of 2020, it is predicted that an astounding 212 billion connected smart items will have been installed globally. For equipment suppliers, Internet service providers, and software developers, the Internet of Things (IoT) presents a substantial business opportunity. By the end of 2020, it is predicted that an astounding 212 billion connected smart items will have been installed globally. This growth signifies the immense potential and widespread integration of IoT technology across various industries and sectors.

Application developers are at the forefront of leveraging the data generated by IoT devices to create innovative applications and services. These developers build software solutions that harness the capabilities of IoT devices, enabling automation, data analysis, remote monitoring, predictive maintenance, and much more. IoT applications span various domains, including smart homes, smart cities, healthcare, agriculture, transportation, and industrial automation.

The Internet of Things (IoT) is a technology that shows promise that aims to connect various smart devices globally, facilitating seamless communication between heterogeneous devices. The A variety of communication technologies with various service requirements have arisen to actualize the vision of the modern IoT in response

Table 1 Contrasting the surveys under scrutiny with the ongoing research efforts

Refs.	IoT	Drone	IoT and drone	Smart cities	Security and privacy	Energy consumption	Blockchain technology	Sensor requirements	Fleet management	Highlighted
[10]	*	*	*	*	*			*	*	The newly presented framework enables the protection of UAVs as "flying" items in a cooperative networked environment by providing a comprehensive IoT architecture
[11]	*	*	*	*						Role of UAVs in various consumer application domains

(continued)

Table 1 (continued)

Refs.	IoT	Drone	IoT and drone	Smart cities	Security and privacy	Energy consumption	Blockchain technology	Sensor requirements	Fleet management	Highlighted
[12]		*		*						Introduces a mechanism named DBCMS (Drone based Covid-19 medical service) designed to safeguard medical personnel vulnerable to Covid-19 infection
[2, 13]	*	*	*	*	*	*	*			The suggested architecture enables the seamless relocation of drones while ensuring the secure transfer of data stored within each zone

(continued)

Table 1 (continued)

Refs.	IoT	Drone	IoT and drone	Smart cities	Security and privacy	Energy consumption	Blockchain technology	Sensor requirements	Fleet management	Highlighted
[6, 14–16]			*	*	*					This extensive study includes a look at the prerequisites, taxonomy, contemporary developments, and difficulties connected with future research trends
[4]		*		*	*	*				Introduces a trusted communication system based on drones for implementation in smart cities

(continued)

Table 1 (continued)

Refs.	IoT	Drone	IoT and drone	Smart cities	Security and privacy	Energy consumption	Blockchain technology	Sensor requirements	Fleet management	Highlighted
[8]	*	*	*	*		*				A drone-based IoTaaS framework is proposed in this paper. This framework aims to optimize deployment, cost, and usage of IoT devices
[3]	*	*	*	*						Introduces DroneTalk, a cutting-edge drone delivery system for mail services, harnessing the power of the Internet of Things (IoT)
Our work	*	*	*	*	*	*	*	*	*	

Note (* In column represent the terms are included in Reference research and blank columns represent reference paper does not include these terms in research) to the rising demand for machine-type communications (MTC). Long-term evolution (LTE) cellular standards, while successful for mobile devices, are not the best choice for low-power, low-data-rate Internet of Things (IoT) devices. As a result, a number of new IoT standards have been created.

In the contemporary landscape, drones are recognized as a revolutionary type of IoT device, significantly influencing the flying IoT sector. With the progress of drone technology, these IoT devices function as sensors, enabling the collection of critical data with strategic value. This data can then be seamlessly transmitted to analytics platforms or cloud-based tools for advanced processing. Moreover, drones find practical applications in conducting aerial inspections of diverse IoT network components, further expanding their capabilities and utility in various contexts.

5 Conclusion and Future Scope

Intelligent entities that can work together intelligently and automatically to improve quality of life, save lives, and sustain resources make up smart cities. Drone technology's introduction has made a substantial contribution to applications in smart cities that are real-time. Drones can be equipped with IoT devices and cooperate with them providing a number of advantages for everyday living and environmental preservation.

The extensive coverage, scope, and purpose of this survey set it apart from other attempts. In particular, it highlights the innovative idea of collaborating with IoT and drones to improve smart city applications. The survey provides a comprehensive overview of recent studies on the application of drones and IoT, highlighting their significance and highlighting prospective use cases for smart cities. It also points out important issues and research findings that must be addressed in order to boost the efficiency of this kind of partnership.

References

1. Amodu OA, Nordin R, Jarray C, Bukar UA, Raja Mahmood RA, Othman M (2023) A survey on the design aspects and opportunities in age-aware UAV-aided data collection for sensor networks and internet of things applications. Drones 7(4):260
2. Al-Fuqaha A, Guizani M, Mohammadi M, Aledhari M, Ayyash M (2015) Internet of things: a survey on enabling technologies, protocols, and applications. IEEE Commun Surv Tutor 17(4):2347–2376
3. Chen KW, Xie MR, Chen YM, Chu TT, Lin YB (2022) DroneTalk: an Internet-of-Things-based drone system for last-mile drone delivery. IEEE Trans Intell Transp Syst 23(9):15204–15217
4. Menouar H, Guvenc I, Akkaya K, Uluagac AS, Kadri A, Tuncer A (2017) UAV-enabled intelligent transportation systems for the smart city: applications and challenges. IEEE Commun Mag 55(3):22–28

5. Yazdinejad A, Parizi RM, Dehghantanha A, Karimipour H, Srivastava G, Aledhari M (2020) Enabling drones in the internet of things with decentralized blockchain-based security. IEEE Internet Things J 8(8):6406–6415
6. Abdelmaboud A (2021) The internet of drones: requirements, taxonomy, recent advances, and challenges of research trends. Sensors 21(17):5718
7. Lai KT, Chung YT, Su JJ, Lai CH, Huang YH (2022) AI wings: an AIoT drone system for commanding ardupilot UAVs. IEEE Syst J
8. Hoque MA, Hossain M, Noor S, Islam SMR, Hasan R (2021) IoTaaS: drone based internet of things as a service framework for smart cities. IEEE Internet Things J
9. Alsamhi SH, Ma O, Ansari MS, Almalki FA (2019) Survey on collaborative smart drones and internet of things for improving smartness of smart cities. IEEE Access 7:128125–128152
10. Lagkas T, Argyriou V, Bibi S, Sarigiannidis P (2018) UAV IoT framework views and challenges: towards protecting drones as "things." Sensors 18(11):4015
11. Vashist S, Jain S (2019) Location-aware network of drones for consumer applications: supporting efficient management between multiple drones. IEEE Consum Electron Mag 8(3):68–73
12. Angurala M, Bala M, Bamber SS, Kaur R, Singh P (2020) An internet of things assisted drone based approach to reduce rapid spread of COVID-19. J Saf Sci Resil 1(1):31–35
13. Qureshi MJ, Kutty AV, Giri J, Dapke A, Giri P, Chadge RB, Sunheriya N (2023) Public acceptance of drones: an approach. CRC Press, pp 493–503. https://doi.org/10.1201/9781003358596-55
14. Khan SWA, Dapke A, Giri J, Sunheriya N (2023) A comparative study of a plane NACA2412 airfoil with a NACA2412 airfoil attached with gurney flap using CFD. CRC Press, pp 1392–1402. https://doi.org/10.1201/9781003370628-74
15. Mahakalkar SG, Giri JP, Sunheriya N, Lande P, Chdge R, Mahatme C, Sathish T (2024) Propulsion effectiveness using propeller selection, ducted propellers and tilting mechanism in a quadrotor drone. Lecture notes in mechanical engineering, pp 399–410. https://doi.org/10.1007/978-981-97-3173-2_26
16. Sunheriya N, Mahakalkar SG, Chadge RB, Lande P, Giri JP, Bagde A (2022). Experimental analysis of twin blade contra-rotating wind turbine. https://doi.org/10.1109/icetems56252.2022.10093565

Contingency Assessment in Power System Having Wind Energy Resource Integration with IEEE 14 Bus System

Jayprakash Giri, Neeraj Kumar Mishra, and Ashwani Tapde

Abstract Contingency evaluation plays a crucial role in assessing and mitigating potential disruptions across various systems. This study enhances our understanding of critical variables and their impact, aiming to develop strategies to minimize adverse effects. Using an IEEE 14 bus system integrated with wind power generation, we conducted comprehensive contingency evaluations to assess system performance and reliability under different scenarios. By simulating diverse contingencies, we analyzed their potential impacts on system stability, security, and overall operation. Performance indices accurately predicted significant line contingencies, identifying critical scenarios likely to exceed system thresholds. Our focus included post-contingency studies on key system factors such as MW flows and bus voltages, aiming to enhance rating and contingency selection processes. Specifically, the performance indixes for (PIp & PIv) active power and reactive power were evaluated, and an overall performance index (OPI) was derived using the DC power flow method. The study identified the line between buses 9 and 14 as particularly sensitive to contingency issues within the examined 14 bus system.

Keywords Contingency evaluation · IEEE 14 bus system · Wind power integration · Performance indices · System reliability

1 Introduction

Providing electricity at an affordable price while keeping an acceptable standard of supply reliability and quality is the basic purpose of an electrical power system. Security and sufficiency are important factors to take into consideration while evaluating a

---

J. Giri (✉) · N. K. Mishra
Electrical Engineering, NIMS University, Jaipur, Rajasthan, India
e-mail: jayprakash.giri@powergrid.in

A. Tapde
TIEIT, Bhopal, Madhya Pradesh, India

A. Kumar et al. (eds.), *Advances in Mechatronics Systems*, Emerging Trends in Mechatronics, https://doi.org/10.1007/978-981-97-8472-1_9

power system's reliability. The term "adequacy" refers to the potential of mass generation and transmission for continuously satisfying the electricity and power demand. The power system's capacity to withstand disruptions without going beyond operational bounds is measured by security, on the contrary. Power system security in deregulated and competitive market situations [1] encompasses more than just uninterrupted power supply. In order to ensure that the system continues to operate even though there is a component failure situation, strategies need to be put into place. Power system failures usually involve transmission lines and producing units. Failures in transmission lines can impact power flows and bus voltages on other lines. When a generator in a power system stops working, it affects the flow of both active and reactive power in addition to the operating parameters of other generators and transmission lines, such as system frequency [2].

Due to the increased use of sustainable energy sources, the growth of demand response initiatives, and the effects of electricity market liberalization, modern power networks are becoming more and more susceptible. The device's controllability has reduced due to this liberalization, forcing electricity systems to perform close to their operating limits. Due to global emission reduction regulations, the secure and sustainable functioning of power systems has turned out to be an essential issue in the energy sector. Power system security is an essential feature of a network that ensures uninterrupted functioning in normal situations, even in the event of an emergency [3]. Blackouts are included under the category of contingencies. Typhoons and other extreme weather events are the primary cause of most blackouts, according to evaluation of their origins [4].

To assure the safe delivery of power to consumers, voltage instability ought to be addressed in the electrical system. Robust power transmission infrastructure is desperately needed because the modern electrical power system is unable to keep up with the constant growth in electric load. New transmission line installation, however, presents difficulties from an economic and environmental standpoint. The increasing integration of renewable energies exacerbates this problem even more. The network's largest and most urgent trouble is the opportunity of transmission lines being overloaded to the point where a massive voltage drop occurs. Voltage collapse can happen due to this state getting worse because of overload on the lines. The line is in a crucial state in this case, which means that even little disruptions might cause the system to collapse [5–8].

N-1 contingency analysis is one among many responsibilities that energy management systems (EMSs) perform, and they may be essential to the supervision of power system operations. When a single power system component, such as a transmission line, power transformer, or generating unit, goes out of service, the consequences of the working circumstances following the removal of that specific component of the overall system have been evaluated by N-1 contingency analysis. The evaluated branch currents are frequently compared to permissible deviations from nominal values, in the process of N-1 contingency analysis. This assessment is done in tandem with the transmission lines' current transmitting capability, the transformer branches' peak loading, and the voltage fluctuations at network nodes. Since many grid regulations require compliance with the aforementioned N-1 criterion, this type of study is essential not only for continuing system operations but also for planning and development. Comprehensive simulations, statistical analysis, sensitivity factors, and overall performance measures are all used in traditional contingency evaluation [9].

Analyzing how probable component failures might affect the power system's overall behavior is vital when it comes to system security. While doing a contingency study, one must examine the consequences of a line outage on the system's overall functionality, while in a contingency evaluation, the security of the system is examined in light of several advised scenarios. A comprehensive analysis of contingencies is crucial to power system planning, operation, and control. Transient phenomena in the power systems' electrical and electromechanical states can result from outages or modifications of their independent properties. Analysis of all workable scenarios needs to be carried out quickly in large power systems with a view to provide control in the less amount of time allotted for corrective action. While applying a diagnostic test to a list of potential contingency conditions comprising multiple system outages is how contingency analysis works. Cases that may lead to reactive power, voltage, and line flow properties being violated need to be diagnosed and ranked according to severity for the purpose of receiving more comprehensive investigation. Contingency analysis comprises of three principal divisions they are; definition, selection, and evaluation [10].

2 Research Contributions

A research by Woldesemayat et al. [11] asserts that contingency problems had been tested in the 400 kV Ethiopian electric energy (EEP) system. The analysis suggests the usage of the gray Wolf An efficient approach to figuring out an Interline Power Flow Controller's (IPFC) ideal functionality. The use of IPFCs improves device security under N-1 contingency eventualities and is exceptionally advantageous. This improvement is indicated by reduced line losses in addition to a reduction inside the Line Severity Index (LSI) and Bus Voltage Severity Index (BVSI). The Gelan to Holeta and Gelan to Sebeta branches are the ideal locations to establish IPFCs, according to the research. Specific parameters are also established in order to enhance machine stability and overall performance.

Abedin et al. [12] highlight the significance of contingency assessment in power system safety, considering both internal and external failures. Their article explored more than one outage scenario in the deregulated environment, highlighting the pressure on the transmission network because of rising load demand. Making use of the Matpower software program, the study examines severity indices for branch failures, generator outages, MVA, and voltage violations. To keep the electrical machine safe, it is crucial to keep an eye on crucial events and branch outages 28, generator outages 23, and generator 22.

In order to forecast the system's severity under various conditions, both with or without a unified power flow controller, Angadi et al. [13] introduce the LVSI (Line Voltage Stability Index) and suggest combining machine learning with data mining approaches. During the contingency assessment stage, there produced a significant amount of data. To properly evaluate severity, it is imperative to increase approaches for converting this data into useful system values. In order to verify the severity of the line, data from simulations with exceptional load situations are used in the evaluation. Data mining and machine learning techniques are used in this procedure. In doing so, the line voltage stability index of the IEEE 30 Bus system has been computed.

Al-Anbarri [14] reported that the key contingencies in a high voltage power system are ranked in this study by taking the usage of an algorithm. The technique used in this presentation is based totally on a hybrid weighted performance index. The hybrid performance index detects bus voltage deviation outside of the permitted limits and overloaded transmission lines. In order to calculate the hybrid performance index, a linear method is used. On the IEEE 24 bus reliability test system, the proposed technique is used. The outcomes of the proposed algorithm and those of the entire AC load flow iterative technique exhibit a good degree of agreement.

The primary objective of the work by Ávila et al. [15] is to examine various power system contingencies using a unique methodology. A different approach based on an algorithm is used in the first stage of this methodology to develop the bus impedance matrix step by step from the system data. On this basis, neither matrix inversions nor transformations will be necessary. Contingencies will be regarded as additional elements, including punctual changes to the bus impedance matrix. The currents utilized come from the base case and are used to obtain the associated voltages on the system, yielding approximative results. In this study, IEEE 30 bus along with IEEE 14 bus systems methodology is examined.

According to Schäfer et al. [16], power system design and real operation both require a quick evaluation of the single contingency policy. Time series data-based grid planning techniques, for an instance, rely on quickly determining loadings during simulated outages. Power system planning strategies that involve multiple power flow computations also rely on this methodology. The most effective prediction of real power flows can be accomplished with strategies like the line outage distribution factor (LODF) matrix and other widely used approximation processes, which have a confined degree of precision. The researchers carry out contingency assessments with artificial neural networks to improve precision and forecast additional power system factors. A whole year's worth of time series simulation results is applied to train deep feedforward network topologies. Comprehensive analysis using a real 110 kV sub-transmission grid in Karlsruhe, Germany, is required for the actual implementation. Forecasts for residual bus voltages and line loadings are produced later. Additional validation of this technique's efficacy is provided by experiments conducted on the IEEE 57 bus network and the CIGRE15 bus medium voltage grid. Compared to the LODF technique, which has an error rate of 18.6%, the prediction errors for each test grid are significantly lower at 0.5%, a major improvement. Additionally, the forecast durations are significantly reduced (10 s vs. 1861s) in comparison to AC power flow calculations (Table 1).

Table 1 Power system contingency assessment: a comparative analysis

Research study	Research objective	Outcomes	Methodology adopted	Numerical values
Woldesemayat et al. [11]	Optimize the function for an interline power flow controller in the EEP system	IPFCs improve system security under N-1 contingency scenarios; reduced line losses and improvements in LSI and BVSI	Gray Wolf Optimization technique; simulation on specific power system branches (Gelan to Holeta, Gelan to Sebeta)	Reduction in Line Severity Index (LSI) and Bus Voltage Severity Index (BVSI)
Abedin et al. [12]	Highlight the importance of contingency assessment in power system safety	Analysis of multiple outage scenarios in a deregulated environment; stress on transmission due to rising load demand	Usage of Matpower software to examine severity indices for various outages and voltage violations	Branch outages 28, generator outages 23, and generator 22 analyzed
Angadi et al. [13]	Forecast system severity under various conditions using UPFCs	Produced a significant amount of data during contingency assessment; computed LVSI for the IEEE 30 Bus system	Combining machine learning with data mining approaches; utilization of data from simulations with exceptional load situations	Computed Line Voltage Stability Index (LVSI) for IEEE 30 bus system
Al-Anbarri [14]	Rank key contingencies in a high voltage power system	Good agreement between the outcomes of the proposed algorithm and the entire AC load flow iterative technique	Usage of a hybrid weighted performance index in a linear method; applied on the IEEE 24 bus reliability test system	Applied on IEEE 24 bus system; comparison with AC load flow results
Ávila et al. [15]	Examine various power system contingencies using a unique methodology	Development of bus impedance matrix step by step from system data without matrix inversions or transformations	Algorithm-based approach to develop bus impedance matrix; examination of IEEE 30 and IEEE 14 bus network	Methodology for the IEEE 14 and IEEE 30 bus networks was investigated

(continued)

Table 1 (continued)

Research study	Research objective	Outcomes	Methodology adopted	Numerical values
Schäfer et al. [16]	Improve the precision of contingency assessment in power system design and real operation using neural networks	Significantly lower prediction errors and reduced forecast durations compared to LODF; validation on multiple bus networks	Artificial neural networks applied to time series data; training of deep feedforward network topologies; application on real grids	Prediction errors reduced to 0.5%, forecast durations from 1861 to 10s
Sunheriya et al. [17]; Khan et al. [18]	Improvement in wind turbine blade design	Enhanced output power	Numerical Simulations	The contra rotating wind turbine generated 25% more power than the single-rotor turbine from the same wind stream

3 IEEE 14 Bus System Design with Wind Energy Resource

Figure 1 demonstrates the IEEE 14 bus system structure, which comprises 9 load buses, 3 generator buses, and 1 slack bus. Furthermore, the system has 3 synchronous compensators that are implemented just to compensate for reactive power and the line data is presented in Table 2, whereas bus data is presented in Table 3. Particularly, DC power flow analysis is used to determine the active power flow across each transmission line under base case loading conditions. The system is modified by the integration of 1 MW wind energy system at bus 2 as another generating unit.

The wind turbine's aerodynamic power output can be stated as:

$$P_{Turbine} = \frac{1}{2}\rho A C_p(\lambda, \beta) v^3 \quad (1)$$

In this equation, v stands for wind speed (in m/s), CP for coefficient of power conversion, A for the area covered by rotor blades (in m^2), and, ρ for air density (generally 1.225 kg/m^3).

The tip-speed ratio can be depicted as:

$$\lambda = \frac{\omega_m R}{v} \quad (2)$$

The rotor radius (in meters) is denoted by R, whereas the rotor angular velocity (in rad/sec) is represented by ω_m.

The mechanical torque output $m\ T$ of the wind turbine can be written as follows:

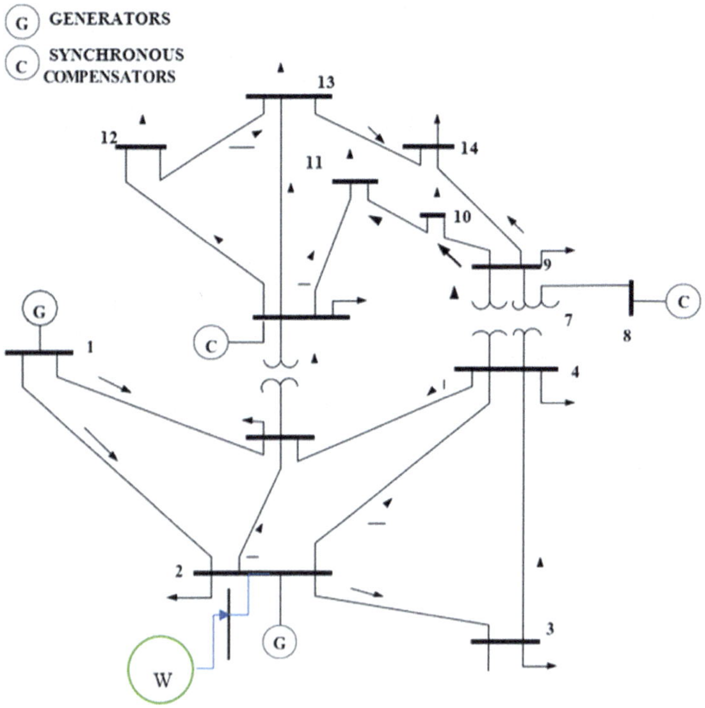

Fig. 1 IEEE 14 bus system using wind energy as a generating source for contingency analysis

$$T_m = \frac{1}{2}\rho A C_p(\lambda, \beta) v^3 \frac{1}{\omega_m} \tag{3}$$

The (β) blade pitch angle (in degrees) and (λ) tip speed ratio's non-linear function is the power coefficient.

Then Power output is given by

$$P_{Turbine} = \frac{1}{2}\rho A C_{p_{\max}} v^3 n \tag{4}$$

The power coefficient, or C_P, is modeled using a general equation that is based on the modeling turbine characteristics given in (1) to (4).

$$C_p = \frac{1}{2}\left(\frac{116}{\lambda_i} - 0.4\beta - 5\right) e^{-\left(\frac{21}{\lambda_i}\right)} \tag{5}$$

Maximum Power Point Tracking (MPPT) is a term used to describe a wind generator's exact location on the power curve that indicates the ideal output power for different wind speeds. As a result, the turbine rotor operates at a variable speed

Table 2 Representing IEEE 14 bus system

Line number	MVA rating	To bus	Half-line charging susceptance (*p.u.*)	Line impedance (*p.u.*)		From bus
				Reactance	Resistance	
1	120	2	0.02640	0.05917	0.01938	2
2	65	5	0.02190	0.22304	0.05403	5
3	36	3	0.01870	0.19797	0.04699	3
4	65	4	0.02460	0.17632	0.05811	4
5	50	5	0.01700	0.17388	0.05695	5
6	65	4	0.01730	0.17103	0.06701	4
7	45	5	0.00640	0.04211	0.01335	5
8	55	7	0	0.20912	0	7
9	32	9	0	0.55618	0	9
10	45	6	0	0.25202	0	6
11	18	11	0	0.1989	0.09498	11
12	32	12	0	0.25581	0.12291	12
13	32	13	0	0.13027	0.06615	13
14	32	8	0	0.17615	0	8
15	32	9	0	0.11001	0	9
16	32	10	0	0.0845	0.03181	10
17	32	14	0	0.27038	0.12711	14
18	12	11	0	0.19207	0.08205	11
19	12	13	0	0.19988	0.22092	13
20	12	14	0	0.34802	0.17093	14

whilst the wind energy Conversion system (WECS) load is adjusted. This makes it viable to continuously extract the wind's maximum power.

4 Contigency Assessment in IEEE 14 Bus System Integrated with Wind Energy Resource

One way to assess these impacts is by calculating performance indices associated with active and reactive power. The general process for conducting contingency assessment and calculating performance indices in the context of the IEEE 14 bus system integrated with the wind energy system has been depicted in Fig. 2.

Appropriate performance indices that reflect the impact of contingencies on active and reactive power aspects are being evaluated. Examples of performance indices include voltage deviation, line overloads, and generator overloads. This work aims to compute the active as well as PIP and PIV, or reactive power and performance indexes respectively. The contingency selection procedure cannot be carried out without

Table 3 Bus data representation for IEEE 14 bus networks

Bus no.	Bus code	Voltage magnitude	Angle degrees	Load		Generator				Injected MVAR
				MW	MVAR	MW	MVAR	Qmin	Qmax	
14	0	1	0	20.86	7	0	0	0	0	0
13	0	1	0	18.9	8.12	0	0	0	0	0
12	0	1	0	8.54	2.24	0	0	0	0	0
11	0	1	0	4.9	2.52	0	0	0	0	0
10	0	1	0	12.6	8.12	0	0	0	0	0
9	0	1	0	41.3	23.24	0	0	0	0	0
8	2	1.09	0	0	0	0	0	−6	24	0
7	0	1	0	0	0	0	0	0	0	0
6	2	1.07	0	15.68	10.5	0	0	−6	24	0
5	0	1	0	10.64	2.24	0	0	0	0	0
4	0	1	0	66.92	10	0	0	0	0	0
3	2	1.01	0	131.88	26.6	0	0	0	40	0
2	2	1.045	0	0	0	232	0	−40	50	0
1	1	1.06	0	30.38	17.78	40	−40	0	0	0

these indices. The scenarios are then organized in descending order of severity, with the scenario with the greatest performance index cost being the maximum serious. Through the usage of energy flow analysis in the MATLAB environment as depicted in Fig. 3, load flow analysis is utilized to compute these indices. From the list of scenarios, the most serious one is selected once the indices have been determined. The bus voltages and power flows across the whole system are systematically analyzed to gain a comprehensive understanding of the impact of the selected contingency on the operational parameters of the system.

Active Power performance index (PI_P): Calculating the volume of line overloads is made easier through this index.

$$PI_P = \sum_{l}^{N_L} \left(\frac{W}{2n}\right)\left(\frac{P_l}{P_l^{max}}\right)^{2n} \tag{6}$$

where,

Number of system lines is indicated by N_L.

A real non-negative weighting factor denoted by W is specified, with a value of (=1).

The MW potential of line l is indicated as P_l^{max}.

The power flow in MW of line l is represented as P_l.

The exponent of the penalty function is denoted by n, with a value of (=1).

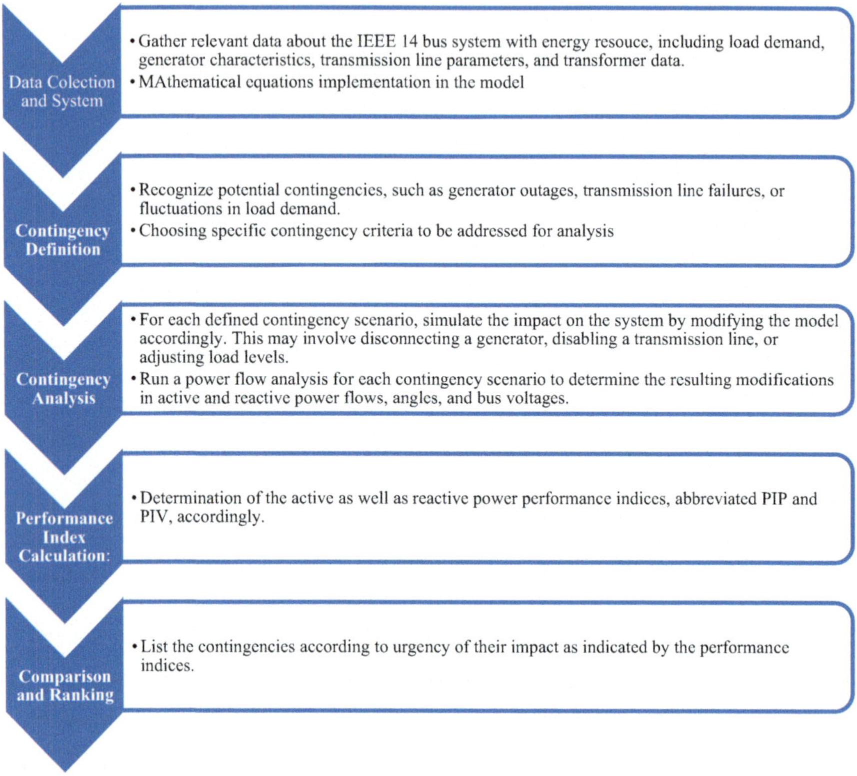

Fig. 2 Contingency assessment process for IEEE 14 bus system with wind energy integration

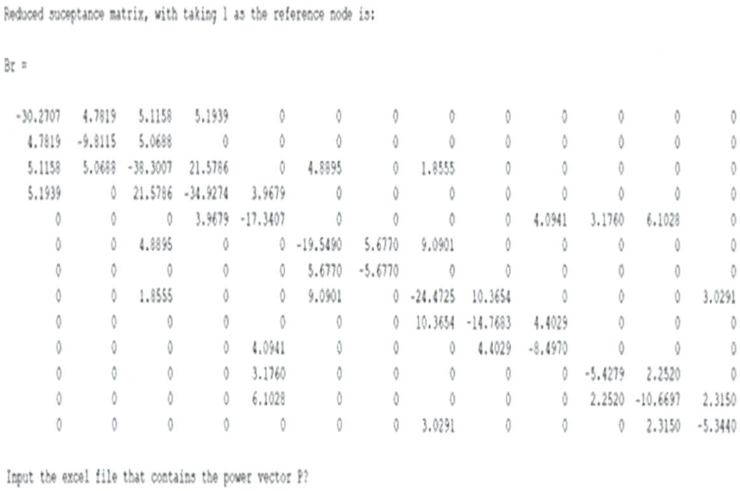
Reduced suceptance matrix, with taking 1 as the reference node is:

Br =

-30.2707	4.7819	5.1158	5.1939	0	0	0	0	0	0	0	0	0
4.7819	-9.8115	5.0688	0	0	0	0	0	0	0	0	0	0
5.1158	5.0688	-38.3007	21.5786	0	4.8895	0	1.8555	0	0	0	0	0
5.1939	0	21.5786	-34.9274	3.9679	0	0	0	0	0	0	0	0
0	0	0	3.9679	-17.3407	0	0	0	0	4.0941	3.1760	6.1028	0
0	0	4.8895	0	0	-19.5490	5.6770	9.0901	0	0	0	0	0
0	0	0	0	0	5.6770	-5.6770	0	0	0	0	0	0
0	0	1.8555	0	0	9.0901	0	-24.4725	10.3654	0	0	0	3.0291
0	0	0	0	0	0	0	10.3654	-14.7683	4.4029	0	0	0
0	0	0	0	4.0941	0	0	0	4.4029	-8.4970	0	0	0
0	0	0	0	3.1760	0	0	0	0	0	-5.4279	2.2520	0
0	0	0	0	6.1028	0	0	0	0	0	2.2520	-10.6697	2.3150
0	0	0	0	0	0	0	3.0291	0	0	0	2.3150	-5.3440

Input the excel file that contains the power vector P?

Fig. 3 Reduced susceptance matrix using bus 1 as the refence bus

$$P_l^{max} = \left(V_i * V_j\right)/X \tag{7}$$

where,

The reactance of the line joining the *ith bus* and the *jth bus* is X.

Vi, derived from the DC power flow solution, is the voltage at bus *i*th.

The voltage at bus *j*th, or Vj, is determined via the DC power flow solution.

Voltage performance index (PI_V): The indices aids in identifying bus voltage restriction violations.

$$PI_V = \sum_{i=1}^{N_B}\left(\frac{W}{2n}\right)\left(\frac{|V_i| - |V_i^{sp}|}{V_i^{lim}}\right)^{2n} \tag{8}$$

where,

ith bus Voltage magnitude is denoted by $|Vi|$.

The defined (rated) voltage magnitude at the ith bus is denoted by $|Vi^{sp}|$.

The voltage's deviation limit is V_i^{lim}.

N_B is the total number of buses utilized in the system.

The value of the penalty function, denoted by n, is (=1).

W is the actual weighting factor which is non-negative, and its value is (=1).

Reactive power impacts bus voltages generated through the generators; so long as the reactive power remains within allowable bounds, the Reactive power performance Index (PIV) offers information on the severity of anomalous voltages. Conventional AC load flow computation is used to compute bus voltages in emergency scenarios where reactive power is close to or exceeds its limitations. This means that at the generator bus, the voltage is different from what is anticipated. As such, boundaries on the reactive power of generators are included in contingency voltage analysis models. Overall performance indices (OPI) are represented by generalized formula:

$$OPI = \sum_{l}^{N_L}\left(\frac{W}{2n}\right)\left(\frac{P_l}{P_l^{max}}\right)^{2n} + \sum_{i=1}^{N_B}\left(\frac{W}{2n}\right)\left(\frac{|V_i| - |V_i^{sp}|}{V_i^{lim}}\right)^{2n} \tag{9}$$

where,

W stands for the non-negative weighting factor with a value of (=1), that has been used.

N_G stands for the total number of generators in the system.

Q_i^{max} is the highest generation of Q_i, while Q_i is the reactive power generated at the *i*th bus.

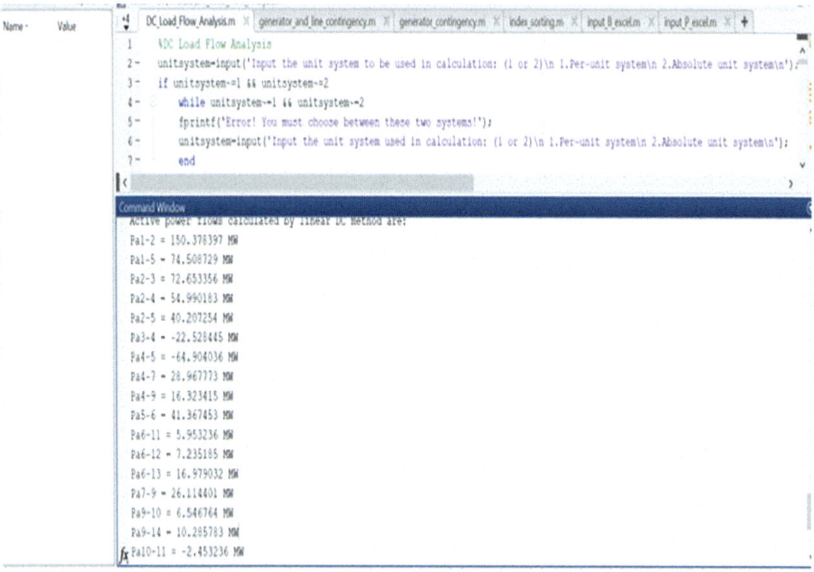

Fig. 4 Power flow prior to emergency in various lines

Since there are twenty transmission strains in the system as shown in Fig. 4, we examine twenty-line contingency conditions by taking one-line loss at a time. Table 4 compiles the performance indexes for each scenario. Line number 17 proves to be highly sensitive because it impacts the entire system, as per the analysis in Table 4, which projects the outage. This accelerated outage PIV rating emphasizes the necessity of paying close attention to this particular line in the course of operations. The performance indices for each line contingency are shown graphically in Fig. 5, where the line numbers that correspond to the outages are indicated on the x-axis and the performance index (PI) value is represented on the y-axis.

The severity of the contingencies has been graded, with 1 being the most critical and 20 being the least severe. Figure 5 suggests the variance of the reactive performance index with respect to respective ranks. The most severe scenario is number 17, which corresponds to the line outage between buses (9–14), consistent with the examination of various PIV values. As an end result, a post-contingency analysis has been carried out about this line outage. Information about the voltage of the system before and after a contingency has been demonstrated in Table 5. The MW flows for rank 1 and rank 2 contingencies, respectively, in both the pre- and post-contingency stages are shown in Tables 6 and 7 (Fig. 6).

Table 4 Performance indicators and contingency ranking for the 14 bus system with DC_load_flow

Outage of the line	PI_P	PI_V	OPI	Ranking
1	1.213	7.0332	8.2462	12
2	1.102	7.2306	8.3326	11
3	1.321	10.0021	11.3231	7
4	1.01	7.669	8.679	10
5	1.062	9.2082	10.2702	8
6	0.968	13.2172	14.1852	2
7	1.006	0.4166	1.1226	20
8	1.4037	1.2553	2.659	17
9	0.9123	11.0045	11.9168	5
10	1.2567	2.6058	3.8625	14
11	1.0153	9.017	10.0323	9
12	1.0165	1.5808	2.5973	18
13	1.101	1.6568	2.7578	16
14	1.007	10.5128	11.5198	6
15	1.109	0.1123	1.2213	19
16	1.012	5.632	6.644	13
17	**1.0186**	**13.3464**	**14.365**	1
18	1.01	11.0012	12.0112	4
19	1.0004	12.1238	13.1242	3
20	1.0054	2.1291	3.1345	15

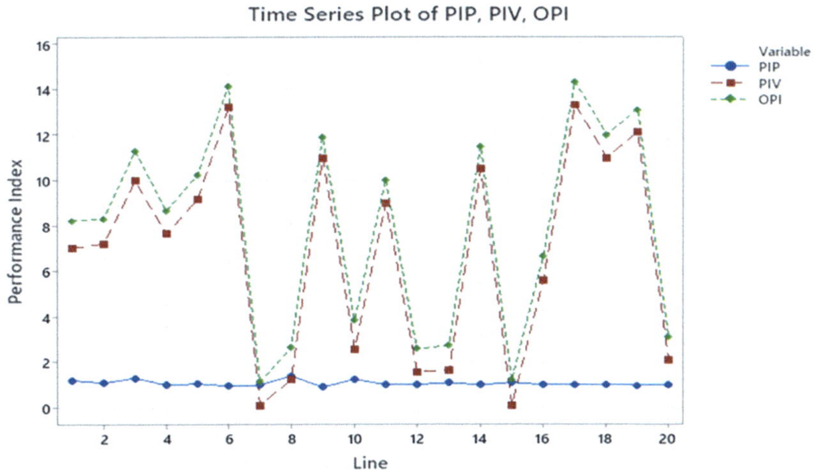

Fig. 5 Performance index of the line in IEEE 14 bus system with wind energy system integration

Table 5 Bus voltage conditions before and after contingencies

Bus number	Pre-contingency voltage (pu)	Post-contingency voltage (pu)
14	0.968	0.970
13	1.001	1.003
12	1.001	1.01
11	1.003	0.989
10	0.998	0.976
9	0.993	0.996
8	1.015	1.016
7	1.007	1.007
6	1.021	1.025
5	1.00	1.009
4	1.002	1.001
3	1.001	1.000
2	1.050	1.051
1	1.074	1.075

Table 6 Active power flow analysis pre and post contingency state rank 1

Line no.	Start bus	End bus	Pre-contingency MW flow	Post contingency MW flow
1	1	2	149.86197	150.1023
2	1	5	74.32415	75.12785
3	2	3	73.28356	72.44137
4	2	4	55.01283	54.01956
5	2	5	40.20725	41.03589
6	3	4	23.12456	23.05789
7	4	5	64.904036	61.45213
8	4	7	28.967773	25.8124
9	4	9	16.127516	15.0159
10	5	6	41.56143	46.2541
11	6	11	6.023586	12.6789
12	6	12	8.023186	7.9434
13	6	13	16.745215	16.1275
14	7	8	0	0
15	7	9	25.94401	24.4127
16	9	10	10.285783	13.01275
17	**9**	**14**	6.546764	0
18	10	11	2.562534	9.0123
19	12	13	1.146289	0.85679
20	13	14	4.623645	2.75812

Table 7 Active power flow analysis pre and post contingency state rank 2

Line no.	Start bus	End bus	Pre-contingency MW flow	Post contingency MW flow
1	1	2	149.86197	155.24963
2	1	5	74.32415	70.015263
3	2	3	73.28356	95.427834
4	2	4	55.01283	44.91235
5	2	5	40.20725	32.55812
6	**3**	**4**	**23.12456**	**0**
7	4	5	64.904036	53.41254
8	4	7	28.967773	29.38493
9	4	9	16.127516	16.802387
10	5	6	41.56143	41.02751
11	6	11	6.023586	5.65124
12	6	12	8.023186	7.253186
13	6	13	16.745215	16.56718
14	7	8	0	0
15	7	9	25.94401	27.00628
16	9	10	10.285783	10.54338
17	9	14	6.546764	6.95361
18	10	11	2.562534	2.125364
19	12	13	1.146289	1.102378
20	13	14	4.623645	4.426358

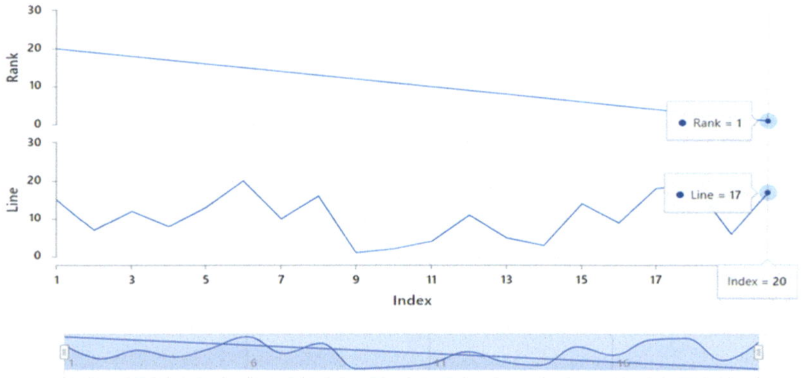

Fig. 6 Ranking plot of the line in the IEEE 14 bus system with wind energy system integration

5 Conclusion

Studying contingency evaluation has yielded valuable records for assessing and handling potential disruptions in a range of systems. Through the identification and examination of potential variables, our consciousness of their significance has grown, which will facilitate the improvement of strategies to reduce their consequences. The present study employed an IEEE 14 bus system and incorporated it with wind electricity as a generating resource. In order to evaluate the system's overall performance and dependability under exclusive scenarios while taking wind power generation into consideration, the IEEE 14 bus system with wind integration underwent a complicated and critical contingency evaluation. For investigating the probable consequences of contingencies on the stability, safety, and overall operation of the system, a variety of scenarios ought to be simulated. Based on the findings, it can be determined that assessing overall performance indices accurately predicts the system's full-size potential line contingencies. The indices with the greatest values suggest a critical scenario having the highest likelihood of surpassing system parameters. As such, among multiple line contingencies, we have determined which is the most serious, and we have carried out a post-contingency study. Particularly, this research concentrates on critical system factors such as MW flows and bus voltages.

Improving the efficacy of rating and contingency selection processes is the aim of this study. In order to accomplish this, the active as well as reactive power performance indices (PIp & PIv), which are essential for contingency evaluation, are evaluated. The voltage performance index and the active power performance index are combined together for getting overall performance index (OPI) by using the DC power flow method. The research, which involved a 14 bus system, found that, within the system under review, the line connecting buses 9 and 14 shows the greatest sensitivity to contingency issues.

References

1. Hailu EA, Nyakoe GN, Muriithi CM (2023) Techniques of power system static security assessment and improvement: a literature survey. Heliyon 9(3):e14524. https://doi.org/10.1016/j.heliyon.2023.e14524
2. Ahmad II, Rawa MJ (2022) Contingency analysis to evaluate power system security and voltage stability for the critical buses. Int J Eng Res Technol (IJERT) 11(02):2278–3181. https://doi.org/10.17577/IJERTV11IS020170
3. Mollaiee A, Taghi Ameli M, Azad S, Nazari-Heris M, Asadi S (2023) Data-driven power system security assessment using high content database during the COVID-19 pandemic. Int J Electr Power Energy Syst 150:109077. ISSN 0142–0615, https://doi.org/10.1016/j.ijepes.2023.109077
4. Panteli M, Pickering C, Wilkinson S, Dawson R, Mancarella P (2017) Power system resilience to extreme weather: fragility modeling, probabilistic impact assessment, and adaptation measures. IEEE Trans Power Syst 32(5):3747–3757. https://doi.org/10.1109/TPWRS.2016.2641463

5. Lee DHA (2016) Voltage stability assessment using equivalent nodal analysis. IEEE Trans Power Syst 31(1):454–463. https://doi.org/10.1109/TPWRS.2015.2402436
6. Aldeen M, Saha S, Alpcan T, Evans RJ (2015) New online voltage stability margins and risk assessment for multi-bus smart power grids. Int J Control 88(7):1338–1352. https://doi.org/10.1080/00207179.2015.1012557
7. Moradi RA, Davarani RZ (2022) Introducing a new index to investigate voltage stability of power systems under actual operating conditions. Int J Electr Power Energy Syst 136:107637. https://doi.org/10.1016/j.ijepes.2021.107637
8. Soyemi A, Misra S, Oluranti J, Ahuja R (2022) Evaluation of voltage stability indices. In: Garg L et al (eds) Information systems and management science (ISMS 2020). Lecture Notes in Networks and Systems, 303, Springer, Cham. https://doi.org/10.1007/978-3-030-86223-7_8
9. Bulat H, Franković D, Vlahinić S (2021) Enhanced contingency analysis—a power system operator tool. Energies 14(4):923 (MDPI AG). https://doi.org/10.3390/en14040923
10. Lekshmi M, Nagaraj MS (2017) A review on electrical power system contingency ranking using artificial intelligence techniques. IOSR J Electr Electron Eng (IOSR-JEEE) 12(4):06–10. https://doi.org/10.9790/1676-1204020610
11. Woldesemayat ML, Tantu AT (2022) Security enhancement of power systems through interline power flow controller (IPFC) under contingency condition: a case study and analysis-EEP 400 kV system. J Electr Comput Eng 1:5897285. https://doi.org/10.1155/2022/5897285
12. Abedin T, Aini N, Pasupuleti J (2022) Application contingency analysis of an IEEE 30 bus system using matpower. J Phys Conf Ser 2319(1):012010 (IOP Publishing). https://doi.org/10.1088/1742-6596/2319/1/012010
13. Angadi RV, Daram SB, Venkataramu PS, Manohar VJ (2022) Data mining and machine learning technique in contingency analysis of power system with UPFC. Distrib Gener Altern Energy J 37(05):1305–1328. https://doi.org/10.13052/dgaej2156-3306.3751
14. Al-Anbarri KA (2020) An approach for contingency ranking analysis of electrical power system. IOP Conf Ser Mater Sci Eng 928(2):022133 (IOP Publishing). https://doi.org/10.1088/1757-899X/928/2/022133
15. Ávila OF, Prado RMB, Silva CMB, Costa VM (2018) Power systems contingency analysis using bus impedance matrix. In: Simposio Brasileiro de Sistemas Eletricos (SBSE), Niteroi, Brazil, pp 1–6. https://doi.org/10.1109/SBSE.2018.8395533
16. Schäfer F, Menke JH, Braun M (2018) Contingency analysis of power systems with artificial neural networks. In: IEEE international conference on communications, control, and computing technologies for smart grids (Martyrdom), Aalborg, Denmark, pp 1–6. https://doi.org/10.1109/SmartGridComm.2018.8587482
17. Sunheriya N, Mahakalkar SG, Chadge RB, Lande P, Giri JP, Bagde A (2022) Experimental analysis of twin blade contra-rotating wind turbine. https://doi.org/10.1109/icetems56252.2022.10093565
18. Khan SWA, Dapke A, Giri J, Sunheriya N (2023) A comparative study of a plane NACA2412 airfoil with a NACA2412 airfoil attached with gurney flap using CFD. CRC Press, pp 1392–1402. https://doi.org/10.1201/9781003370628-74

Mathematical Modelling and Speed Control of SRM Drive

Stella Kurian and Vinita Chellappan

Abstract Switched Reluctance Motor (SRM) is a synchronous AC machine which extracts mechanical power from reluctance torque. This chapter presents the mathematical modelling and small signal modelling of switched reluctance machine. Speed control of Switched reluctance motor drive with PID speed controller employing hysteresis controller and sliding-mode controller for inner current loop is presented. Simulation results of open loop, current control loop and speed control loop including transient response analysis is presented. The results of the transient introduced to variation in various input conditions and load torque is studied and analysed for both current control and speed control loop. The settling time reduces on including the SMC current controller.

Keywords SRM drive · Mathematical modelling · Small signal modelling · Hysteresis controller · PID controller · Transient response · Sliding-mode controller

1 Introduction

SRM is a preferred alternative for EV/HEV industry since it is 30–40% cheaper than the traditional EV motors because of its various advantages including simple and rugged construction, robust and high-speed capability, fault tolerance since each phase is independent and fault in one phase doesn't affect the others, low-production cost since rotor windings and permanent magnets are absent. Extended constant power speed range makes up for the major disadvantages of lower torque density and higher torque ripple in automotive applications [1–3]. SRM requires control circuitry to be operated and recent development in electronic control techniques have

S. Kurian (✉) · V. Chellappan
Government Engineering College Palakkad, APJ Abdul Kalam Technological University, Palakkad, Kerala, India
e-mail: stellakurian31@gmail.com

V. Chellappan
e-mail: vinitac@gecskp.ac.in

A. Kumar et al. (eds.), *Advances in Mechatronics Systems*, Emerging Trends in Mechatronics, https://doi.org/10.1007/978-981-97-8472-1_10

led to evolution of SRM to achieve desired performance [4, 5]. The paper focuses onto design and analyse current control and speed control schemes under varying load and input conditions, and studies effectiveness in reducing the torque ripple of the SRM motor. Section 2 briefs the operation of SRM and Sect. 3 presents the mathematical modelling of the machine. Section 4 explains the speed control scheme briefing the open loop, current control and speed control scheme. Simulation results of open-loop operation, current control loop employing a hysteresis controller and a second-order sliding-mode controller, and their transient analysis are presented in Sects. 5 and 6, respectively. Speed control loop implementation with PID controller with transient conditions is shown in Sect. 7.

2 Switched Reluctance Motor

SRM has a dual salient structure with its projecting stator and rotor poles [6]. When a particular phase is excited, rotor pole tends to move to get aligned with the next stator pole. The variation in air gap proportionally leads to variation in phase inductance of the stator. The actual inductance profile can be divided into three regions as shown in Fig. 1 [7]. Here, L_A, L_B and L_C are the three phase inductances. The phase windings are sequentially excited in the order ABC. The incoming phase turn on angle is denoted by θ_{on1} and the next incoming phase turn on angle is denoted by θ_{on2}. The region between θ_{on1} and θ_1 is the overlapping region and the poles come in alignment in the region between θ_1 and θ_2. Each phase accounts for 120° in electrical angle.

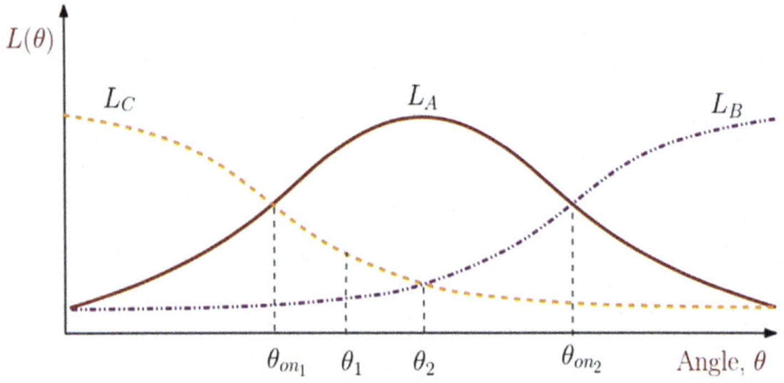

Fig. 1 Inductance profile of SRM [7]

3 Mathematical Modelling

Per-phase equivalent circuit formulation for the SRM is developed by ignoring the mutual coupling among the stator windings [8]. Applied excitation to any stator winding is given by,

$$v_s = R_s \cdot i + \frac{d\lambda_s(\theta, i)}{dt} \tag{1}$$

R_s is the resistance and λ_s is the flux linkage of single winding, wherein the flux linkage of a phase is function of inductance and L_s phase current, i,

$$\lambda_s(\theta, i) = L_s(\theta, i) \cdot i \tag{2}$$

Equation (1) can be rewritten as,

$$v_s = R_s \cdot i + L_s(\theta, i) \cdot \frac{di}{dt} + i \cdot \frac{d\theta}{dt} \cdot \frac{dL_s(\theta, i)}{d\theta} \tag{3}$$

$$v_s = R_s \cdot i + L_s(\theta, i) \cdot \frac{di}{dt} + i \cdot \omega_m \cdot \frac{dL_s(\theta, i)}{d\theta} \tag{4}$$

The first term of Eq. (4), ($R_s \cdot$ i) is the resistive drop, second term, $\left(L_s(\theta, i) \cdot \frac{di}{dt}\right)$ inductive voltage drop and the last term $\left(i \cdot \omega_m \cdot \frac{dL_s(\theta,i)}{d\theta}\right)$ is the induced emf, e. Also,

$$e = K_b \cdot \omega_m \tag{5}$$

where, K_b is the emf constant. From these equations, single phase equivalent diagram is developed. See Fig. 2 [9].

Now, the power developed per phase is given by,

$$p = e \cdot i = i^2 \cdot \omega_m \cdot \frac{dL_s(\theta, i)}{d\theta} + L_s(\theta, i) \cdot i \cdot \frac{di}{dt} \tag{6}$$

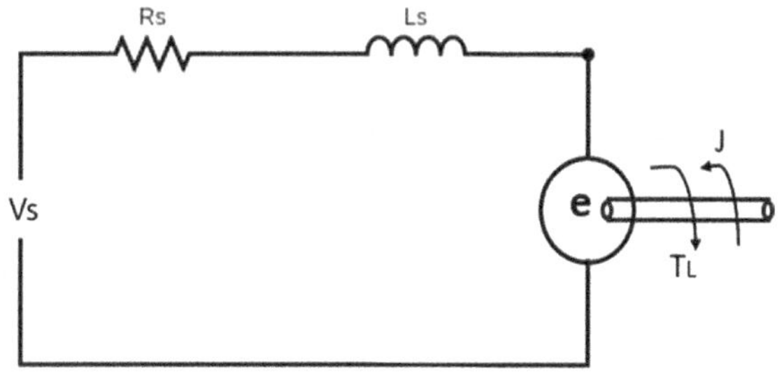

Fig. 2 Equivalent diagram [9]

Energy stored in magnetic field [7] is,

$$W_e = \frac{1}{2} L_s(\theta, i) \cdot i^2 \tag{7}$$

Power due to variation in magnetic field is given by,

$$\frac{d}{dt}\left(\frac{1}{2} L_s(\theta, i) \cdot i^2\right) = L_s(\theta, i) \cdot i \cdot \frac{di}{dt} + \frac{1}{2} i^2 \cdot \frac{dL_s(\theta, i)}{dt} \tag{8}$$

Resultant air gap power or mechanical power developed, P_m is the difference between electrical power developed and power due to variation in magnetic field. So, from Eqs. (7) and (8),

$$P_m = \frac{1}{2} i^2 \cdot \frac{dL_s(\theta, i)}{d\theta} \frac{d\theta}{dt} \tag{9}$$

$$P_m = \omega_m \cdot T_e \tag{10}$$

Electromagnetic torque can then be equated as,

$$T_e = \frac{1}{2} i^2 \cdot \frac{dL_s(\theta, i)}{d\theta} \tag{11}$$

Mechanical equation of the motor is,

$$\begin{aligned} T_e &= J \frac{d\omega_m}{dt} + B\omega_m + T_L \\ T_e - T_L &= J \frac{d\omega_m}{dt} + B\omega_m \end{aligned} \tag{12}$$

$$\frac{1}{2} i^2 \cdot \frac{dL_s(\theta, i)}{d\theta} - T_L = J \frac{d\omega_m}{dt} + B\omega_m \tag{13}$$

Small Signal Modelling or Linearized modelling of SRM

Small signal linearization around an operating point is done by considering a non-saturating motor with a constant stator current. Also, the self-inductance of each phase is assumed to depend upon the rotor position only. So, $L_s(\theta, i)$ becomes $L(\theta)$. Rotor dependency on current is ignored by selecting $L(\theta)$ as the mean value of maximum and minimum phase inductance [10]. Perturbations near the steady state operating point are added for the following states as,

$$\begin{aligned} v_s &= v_{so} + \delta v_s \\ \omega_m &= \omega_{mo} + \delta\omega_m \end{aligned}$$

$$
\begin{aligned}
i &= i_o + \delta i \\
T_L &= T_{Lo} + \delta T_L
\end{aligned} \tag{14}
$$

where, subscript 'o' denotes the operating point and δ is the incremental deviation near the operating point. Variation is phase current is hence expressed from phase voltage equation, Eq. (4) as,

$$
\frac{di}{dt} = \frac{v_s}{L} - R_s \cdot i - \frac{1}{L}\frac{dL}{d\theta} \cdot \omega_m \cdot i \tag{15}
$$

Substituting perturbated values,

$$
\frac{d(i_o + \delta i)}{dt} = \frac{(v_{so} + \delta v_s)}{L} - \frac{R_s}{L}(i_o + \delta i) - \frac{1}{L}\frac{dL}{d\theta}(\omega_{mo} + \delta\omega_m)(i_o + \delta i)
$$

Applying steady state values in Eq. (15) leads to,

$$
\frac{d(\delta i)}{dt} = \frac{\delta v_s}{L} - R_s(\delta i) - \frac{1}{L}\frac{dL}{d\theta}(\omega_{mo})(\delta i) - \frac{1}{L}\frac{dL}{d\theta}(\delta\omega_m)(i_o) \tag{16}
$$

Let the equivalent per phase resistance, R_{eq}, emf constant, K_b and variation in induced emf, δe be,

$$
\begin{aligned}
R_{eq} &= R_s + \omega_{mo} \cdot \frac{dL}{d\theta} \\
K_b &= i_o \cdot \frac{dL}{d\theta} \\
\delta e &= i_o \cdot \delta\omega_m \cdot \frac{dL}{d\theta}
\end{aligned} \tag{17}
$$

Equation (16) becomes

$$
\frac{d}{dt}(\delta i) = \frac{\delta v_s}{L} - \frac{\delta e}{L} - \frac{R_{eq}}{L}(\delta i) \tag{18}
$$

From Eq. (13),

$$
\frac{d\omega_m}{dt} = -\frac{B}{J}(\omega_m) - \frac{1}{J}(T_L) + \frac{1}{2J}i^2\frac{dL(\theta)}{d\theta} \tag{19}
$$

Substituting perturbated values [9] and solving the equations yields,

$$
\begin{aligned}
\frac{d(\omega_{mo} + \delta\omega_m)}{dt} &= \frac{1}{2J}(i_o + \delta i)^2\frac{dL(\theta)}{d\theta} - \frac{B}{J}(\omega_{mo} + \delta\omega_m) - \frac{1}{J}(T_{Lo} + \delta T_L) \\
\frac{d}{dt}(\delta\omega_m) &= \frac{1}{J}(i_o \cdot \delta i)\frac{dL}{d\theta} - \frac{B}{J}(\delta\omega_m) - \frac{1}{J}(\delta T_L)
\end{aligned}
$$

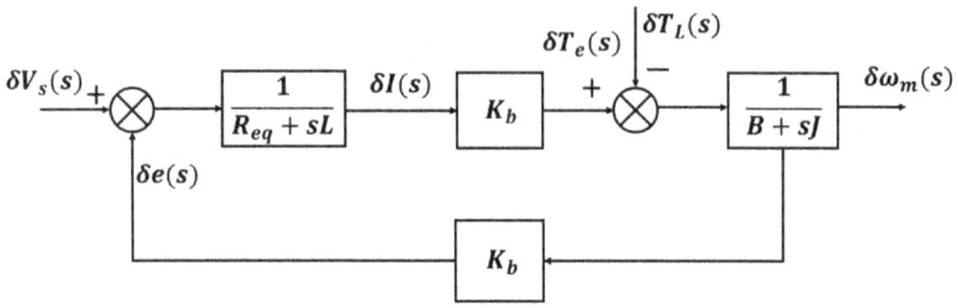

Fig. 3 Linearized model of SRM

$$= \frac{1}{J}(K_b \cdot \delta i) - \frac{B}{J}(\delta\omega_m) - \frac{(\delta T_L)}{J} \tag{20}$$

On applying Laplace transformations,

$$\delta I(s) = \frac{\delta V(s) - \delta e(s)}{R_{eq} + sL} \tag{21}$$

$$\delta\omega_m(s) = \frac{K_b \cdot \delta I(s) - \delta T_L(s)}{Js + B} \tag{22}$$

$$\delta e(s) = K_b \cdot \delta\omega_m(s) \tag{23}$$

From these equations, the linearized model of SRM is developed as shown in Fig. 3.

4 Speed Control Scheme

Speed control scheme with a speed controller, current controller, converter and SRM motor is shown in Fig. 4. The scheme shows the outer speed and inner current control loop. The SRM is excited through a converter from a DC supply which distributes the DC power among the phases in synchronization with position of rotor poles. The rotor position is read from the motor by rotor position sensor [11].

To run the motor in open loop, rotor position signals are operated among the turn-on and turn-off angles and gate pulses for the converter are generated. Here an asymmetric converter is used which operates in three modes, magnetizing, demagnetizing and freewheeling modes in sequence such that positive, negative and zero DC voltage appears across the phase windings respectively [12].

For the inner current control loop, the reference current, (I_{ref}) and the actual stator phase winding current value, (i) fed back from the motor are compared [13]. The error from the comparator gets modulated by the current controller to produce a controlled signal based on the rotor position signals, to obtain the gate pulses for

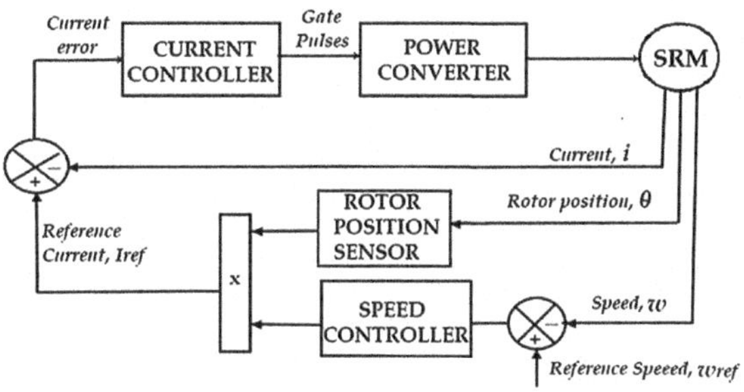

Fig. 4 Closed loop speed control scheme

the switches of converter. Two current controllers are used to compare the results, a hysteresis current controller and a second-order sliding-mode current controller.

Hysteresis current controller (HCC) ensures that the instantaneous current value is bounded within the selected hysteresis band [14]. When the current reaches the upper boundary, the gate pulses go low turning off the phase winding and when the current values reach the lower boundary, the gate pulses become high turning on the switches. The major limitation of this controller is the increase in switching frequency as lower hysteresis band results in lower current ripple [15].

Sliding-mode current controller (SMC) is used with super twisting algorithm as another option for current control to reduce the switching frequency and current ripple. Discontinuous control acts on second derivative of current error, $err_i = (i^* - i)$ [16]. The current error is considered as the sliding surface, s.

$$\frac{ds}{dt} = \frac{di^*}{dt} - \frac{v_s}{L_s} - \frac{(e + R_s i)}{L_s} \tag{24}$$

we set

$$\frac{ds}{dt} = \phi(t, s) + \psi(t, s) v_s \tag{25}$$

with

$$\phi = \frac{(e + R_s i)}{L_s} \text{ and } \psi = -\frac{1}{L_s}$$

with hypothesis as $\frac{di^*}{dt} = 0$ as i^* is constant. Assumptions to ensure the accessibility of the sliding surface:

- variables must be bounded [17].
- there exist positive constants S_0, K_m, K_M, C_0 such that $\forall x \in R^n \ \exists |s(t, x)| \le S_o$.

System satisfies the condition:

$$0 < K_m \le |\psi(t, s)| \le K_M \ |\phi(t, s)| \le C_o$$

Control voltage, v_s is composed of two continuous terms, integral term, U_i and proportional term, U_p,

$$\frac{dU_i}{dt} = \beta \cdot sgn(s) \tag{26}$$

$$U_p = \alpha\sqrt{s} \cdot sgn(s) \tag{27}$$

with $\alpha > 0$ and $\beta > 0$ [18].

Algorithm integrates the polarity of error and the proportional gain is a function of square root of the error, which varies much faster than the error in the vicinity of the command value reducing the settling time [19]. For implementing the outer speed control loop, the error between the measured or actual rotor speed (ω_m) and the reference speed (ω_{ref}) is controlled by the PID speed controller to produce a current reference for the current controller. Reference current is distributed among the three phases corresponding to the rotor position signals read from the motor.

5 Open Loop Operation of SRM

A 60 kW machine energized with 240 V DC supply through the asymmetric converter is simulated in MATLAB simulink. For sequential excitation of the phases through the converter rotor position signals from the motor are necessary [20]. Table 1 shows the machine parameters of SRM chosen for simulation. Figure 5 shows the expanded output waveforms of the open loop operation of SRM drive. Stator phase voltage waveform shows three levels of positive DC voltage, negative DC voltage and zero voltage level coinciding with the magnetizing, demagnetizing and freewheeling mode of the converter.

When the magnetizing mode is turned on the flux develops and then gets reduced when the mode changes to demagnetizing mode. The flux value becomes zero during the freewheeling mode of the converter. Stator current similarly increases when a

Table 1 SRM motor parameters

Parameter	Value
Stator resistance	0.05 Ω
Friction	0.02 Nms
Inertia	0.05 kgm^2
Minimum inductance	0.67 mH
Maximum inductance	23.6 mH
Initial rotor position and speed	0 rad, 0 rad/sec
Maximum flux linkage	0.486 Vs
Maximum current	450 A

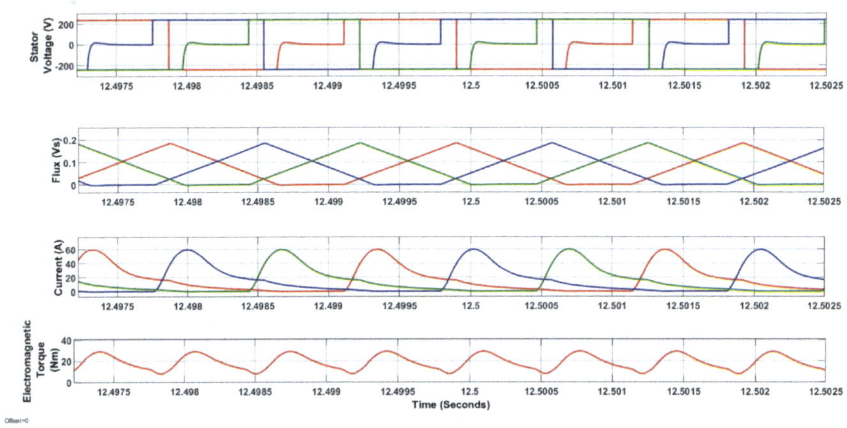

Fig. 5 Output waveforms of open loop SRM drive

particular phase gets turned on. It reaches its peak and is then reduced when the next incoming phase goes to freewheeling mode. i.e. when the voltage across the next incoming phase is zero. The peak current reaches nearly 60 A when an excitation voltage of 240 V is applied. The torque waveform shows a ripple of 9.5 Nm for an applied load torque of 3 Nm. Waveforms depict that there is no control in the current, torque or speed values.

Transient response with load torque change:

Figure 6 shows the speed and load torque waveform when a load torque of 3 Nm is applied at 2.5 s. At no load condition, speed gets settled to 980 rad/s (9360 RPM) nearly at 1.5 s in the open loop operation. When load torque of 3 Nm is applied, the speed reduces to 870 rad/s (8300 RPM) and settles in 1.5 s.

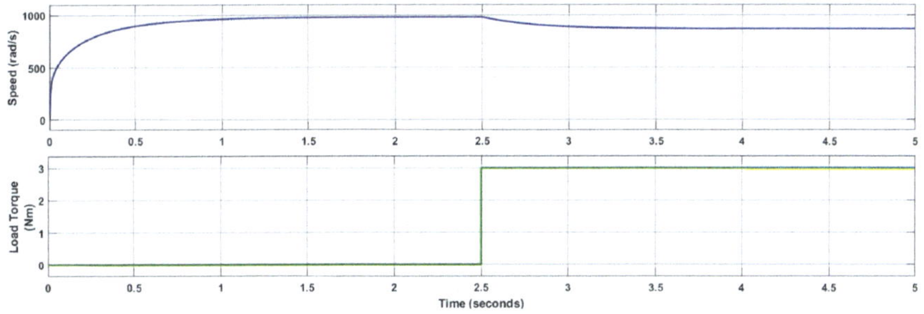

Fig. 6 Speed waveform of open loop SRM drive with load change

6 Current Controlled SRM Drive

In a current controlled SRM drive, current controller and the converter are only implemented as part of the inner current control loop. The current controller operates on the current error between reference and the actual current measured from the machine windings. The controlled signal gets operated with position feedback signals so as to obtain the sequentially excite the phase windings.

Current control with HCC

A hysteresis controller is used with an allowable band of $\pm 10\%$. Simulation results of the inner current loop with a reference peak current of 20 A is shown in Fig. 7 and the expanded waveforms are shown in Fig. 8. Waveforms show that the peak current is limited to 20 A throughout the simulation. Speed gets settled to 6425 RPM in 3 s time. Torque ripple of 6 Nm is observed in the current controlled system.

Transient response with load torque change:

Transient response of the current-controlled drive with sudden load change of 2 Nm at 4 s is depicted in Fig. 9 with a reference peak current of 20 A. It is very well clear from the current waveform that the peak current gets maintained at the reference value of 20 A.

Speed gets reduced since there is no speed control loop from 6425 to 5545 RPM when a load torque of 2Nm is added to the system which gets affected at nearly 7 s and is not shown in the expanded view in Fig. 9. The settling time for the speed

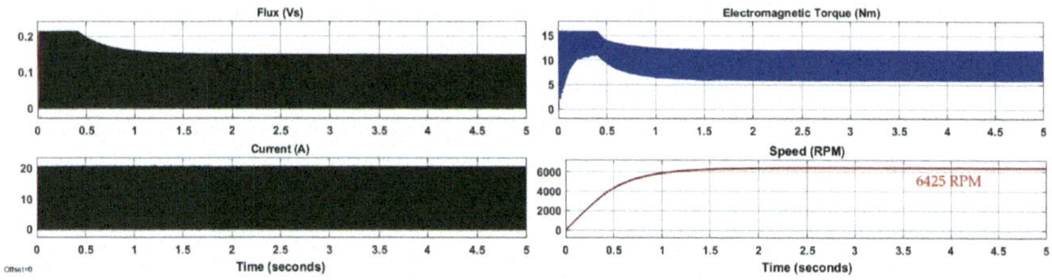

Fig. 7 Output waveforms of HCC current controlled SRM drive

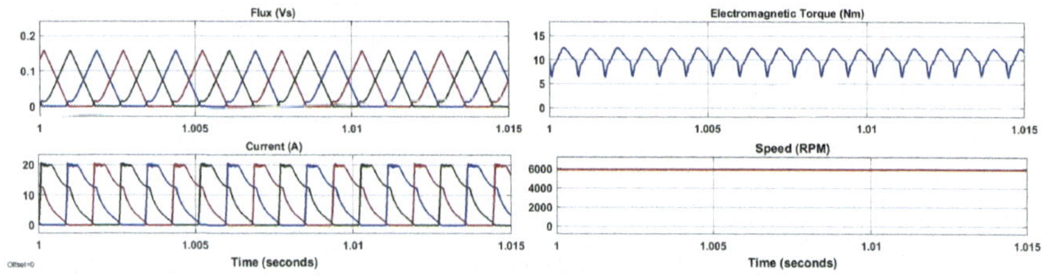

Fig. 8 Expanded output waveforms of HCC current-controlled SRM drive

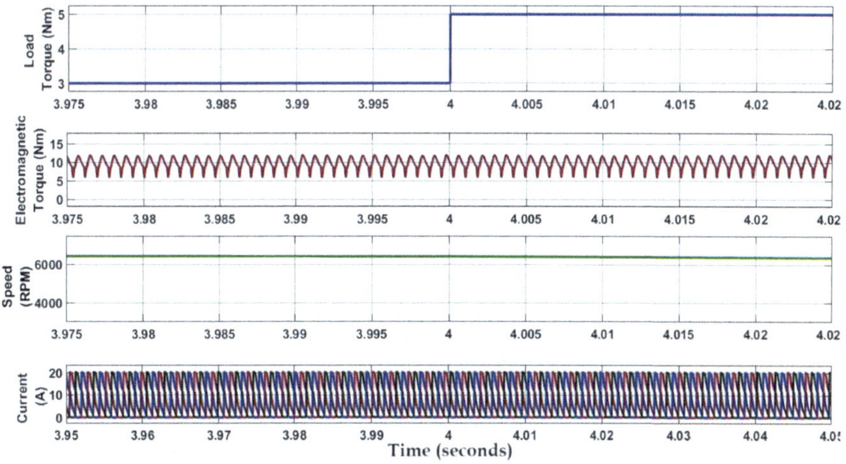

Fig. 9 Waveforms of HCC SRM drive with sudden load change

waveform is nearly 3 s as observed from the simulation results, which is more than the open loop system.

There is no significant increase in the torque ripple value, though the average torque increases.

Transient response with change in reference peak current:

A sudden change from 20 to 30 A is applied to the current controlled drive at 4 s and the response is depicted in Fig. 10 with a load torque of 3 Nm. It is very well clear from the current waveform that the peak current increases to 30 A in a few microseconds. Speed increases from 6425 RPM and settles at 7336 RPM in nearly 2 s and is not shown in the expanded view in Fig. 10. The torque ripple value increases from 9.2 Nm to 10 Nm.

Current control with SMC

Simulation results of the current controlled drive employing SMC with a peak current reference of 20 A is shown in Fig. 11 and the expanded waveforms are shown in Fig. 12. Waveforms shows that the peak current is limited to 20 A throughout the simulation. Speed gets settled to 6388 RPM in 2.6 s time. Torque ripple of 6 Nm is observed in the current controlled system.

Transient response with load torque change:

Transient response of the current controlled drive with sudden load change of 2 Nm at 4 s is presented in Fig. 13 with reference peak current of 20 A. It is very well clear from the current waveform that the peak current gets maintained at the reference value of 20 A. Speed gets reduced since there is no speed control loop from 6388 to 5517 RPM on applying a load torque of 2 Nm and is not shown in the expanded view in the speed waveform. The settling time for the speed waveform is nearly 1.8 s

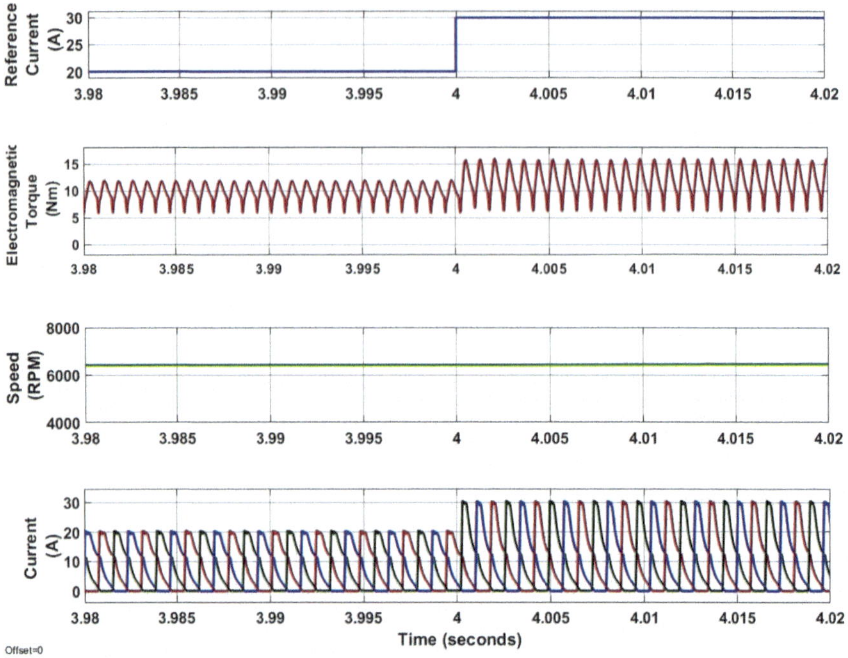

Fig. 10 Waveforms of HCC SRM drive with change in reference current

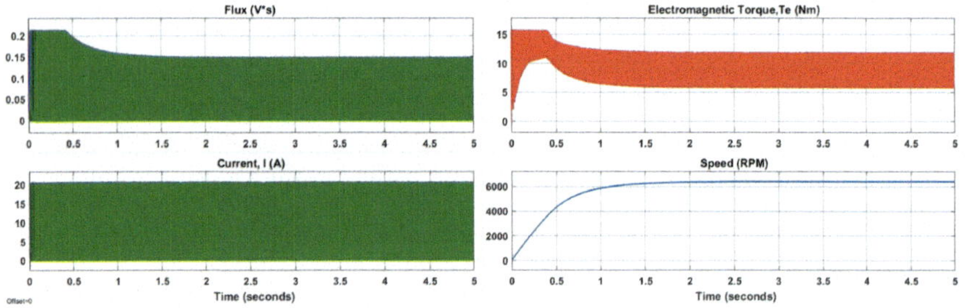

Fig. 11 Output waveforms of SMC current controlled SRM drive

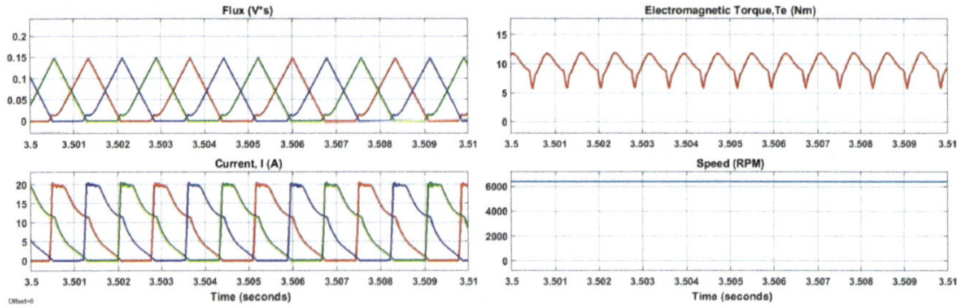

Fig. 12 Expanded output waveforms of SMC current controlled SRM drive

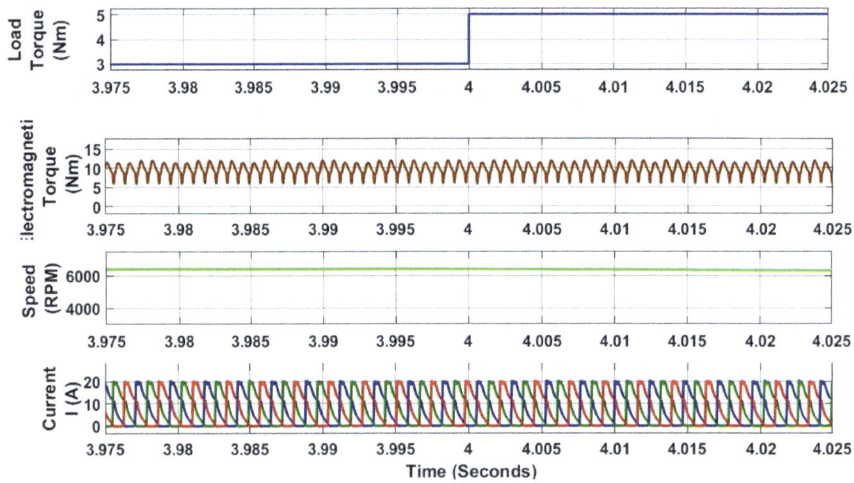

Fig. 13 Waveforms of SMC-controlled SRM drive with sudden load change

as observed from the simulation results, which is more than the open loop system. There is no significant increase in the torque ripple value, though the average torque increases.

Transient response with change in reference current:

A sudden change of current from 20 to 30 A is applied at 4 s and response is depicted in Fig. 14 with a load torque of 3 Nm. It is very well clear from the current waveform that the peak current increases to 30 A in few microseconds. Speed increases from 6338 RPM and settles at 7334 RPM in nearly 1.8 s which is not shown in the expanded view of the speed waveform. The torque ripple value increases by 1.5 Nm.

7 Speed-Controlled SRM Drive

This section deals with the implementation of the complete speed control scheme shown in Fig. 4 with a PID speed controller and current controller. Simulation with both current controllers is presented with transient analysis.

Speed Control with HCC

The PID speed controller is tuned with gain constants $K_i = 0.65$, $K_p = 0.5$ and $K_d = 0.005$ and the hysteresis band of HCC is selected to be $\pm 10\%$. Simulation results of the speed-controlled drive with a speed reference of 2500 RPM are shown in Fig. 15 and the expanded waveforms are shown in Fig. 16. Waveforms show that the speed gets settled to reference speed 2500 RPM in 23 ms time. The settling time was in order of seconds in open loop and current controlled SRM drive. The peak

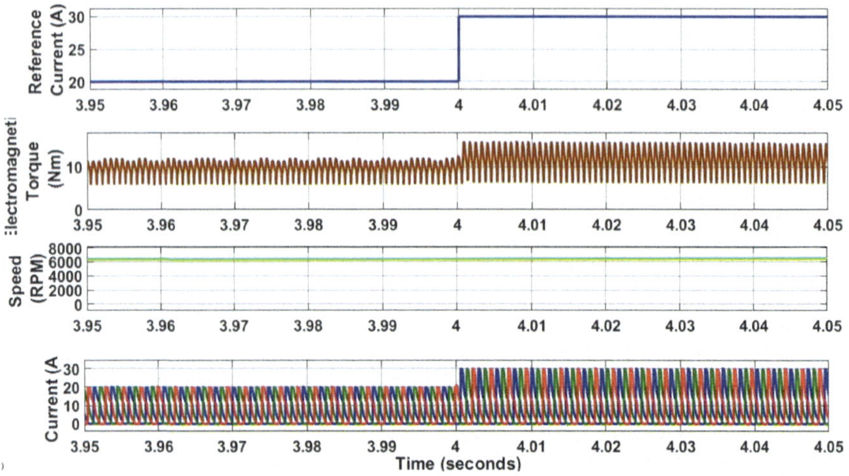

Fig. 14 Waveforms of SMC SRM drive with change in reference current

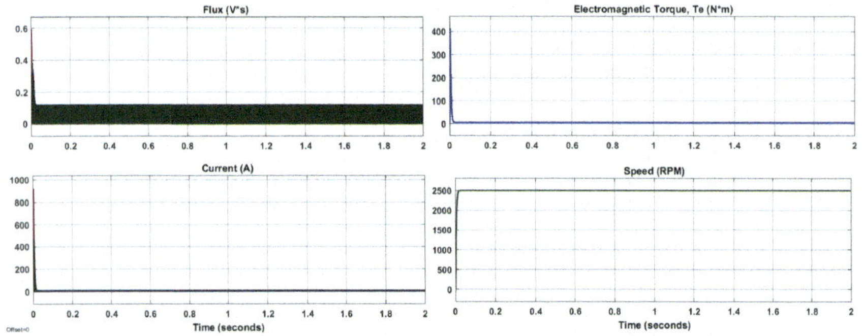

Fig. 15 Output waveforms of speed-controlled HCC SRM drive with reference speed of 2500 RPM

current shows only 9.2 A and the torque ripple of only 2.5 Nm for the same load torque compared to the current controlled system.

Transient response with load torque change:

Transient response of the speed-controlled drive with sudden variation in load torque to 10 Nm from 0.5 s is depicted in Fig. 17 with a reference speed of 2500 RPM. It is seen that the speed is slightly reduced by 20 RPM to 2480 RPM. Speed reduction is regained in 2 s and the speed is maintained at 2500 RPM again. There is no significant increase in the torque ripple value, though the average torque increases.

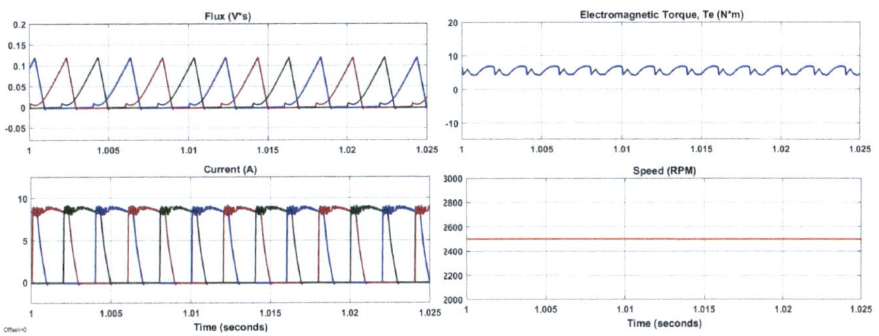

Fig. 16 Expanded output waveforms of speed-controlled HCC SRM drive with reference speed of 2500 RPM

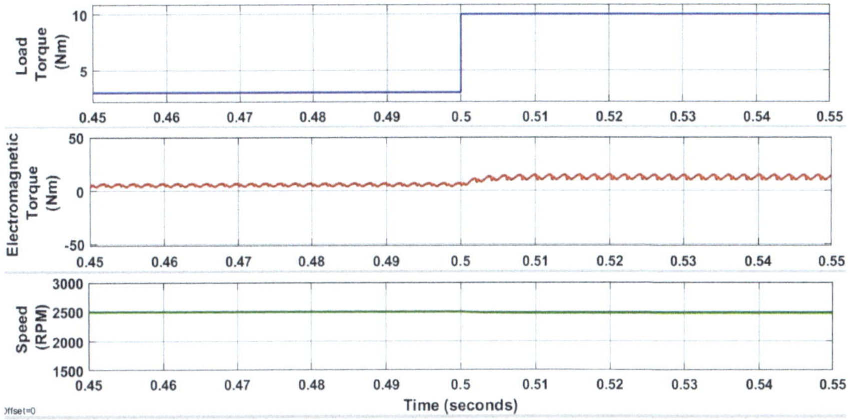

Fig. 17 Waveforms of speed-controlled HCC SRM drive with sudden load change

Transient response with change in reference speed:

Transient response of the speed-controlled HCC drive with sudden change in reference speed value from 2500 to 3000 RPM at 0.5 s is depicted in Fig. 18 with a load torque of 3 Nm. It is very well clear from the transient change in speed that is tracked in 23 ms time. The peak current increases from 9.2 A to 9.7 A and there is no considerable increase in torque ripple.

Speed Control with SMC

The PID controller is tuned with gain constants $K_i = 0.75$, $K_p = 0.5$, and $K_d = 0.005$ and sliding mode controller is used here to implement the speed control. Simulation results of speed-controlled drive with a speed reference of 2500 RPM are shown in Fig. 19 and the expanded waveforms are shown in Fig. 20. Waveforms

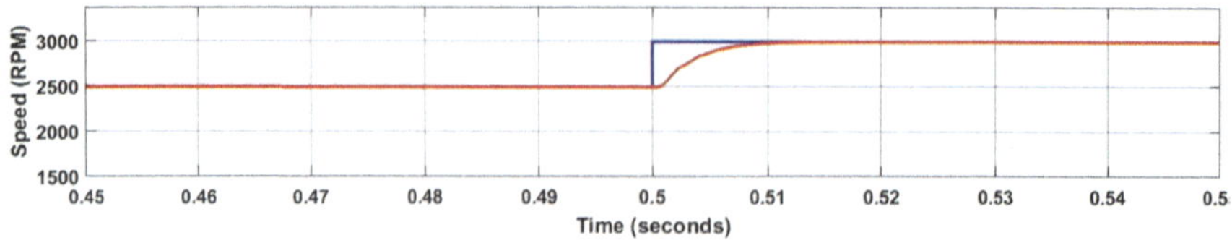

Fig. 18 Speed waveform of speed-controlled HCC SRM drive with sudden change in reference speed

shows that the speed gets settled to reference speed 2500 RPM in 16 ms time. The settling time was in order of seconds in open loop and current controlled SRM drive.

Transient response with load torque change:

Transient response of the speed-controlled drive with sudden load change from 3 to 10 Nm at 0.5 s is depicted in Fig. 21 with a reference speed of 2500 RPM. It is very well clear that the speed is slightly reduced by 20 RPM to 2480 RPM. Speed reduction is regained in 1.8 s and the speed is maintained at 2500 RPM again. There is increase in the torque ripple value from 2 to 5 Nm.

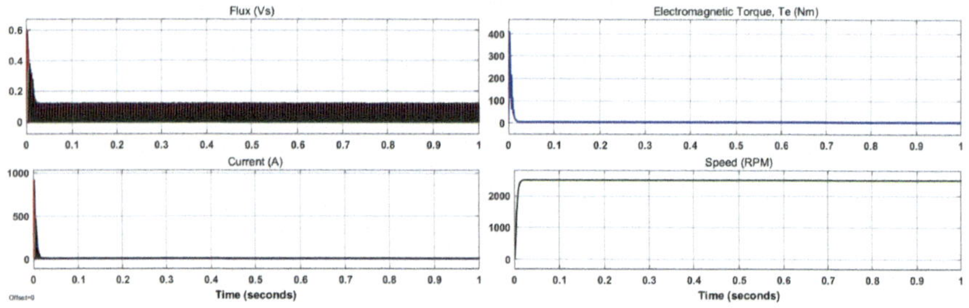

Fig. 19 Output waveforms of speed-controlled SMC SRM drive with reference 2500 RPM

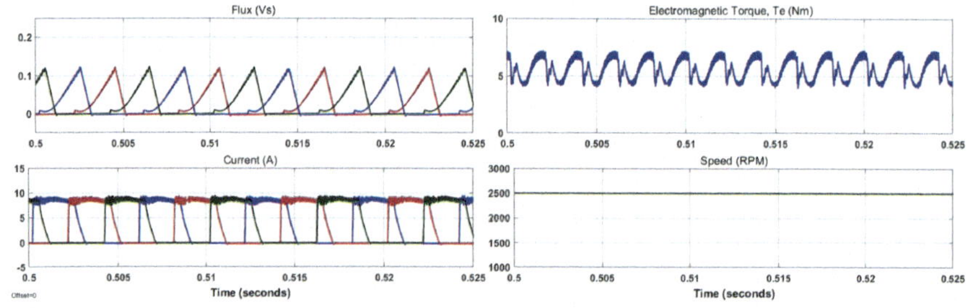

Fig. 20 Expanded output waveforms of speed-controlled SMC SRM drive with reference speed of 2500 RPM

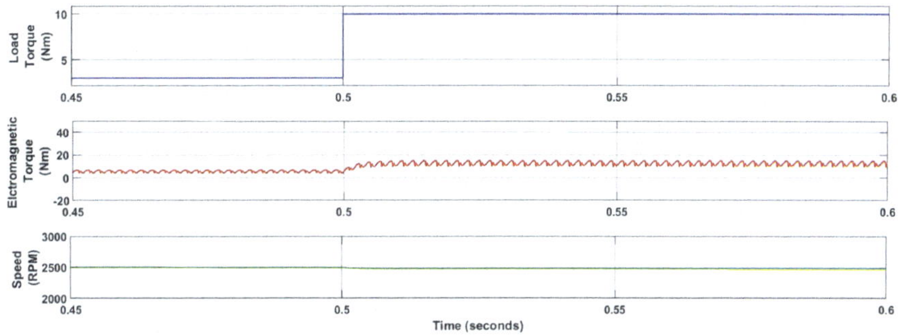

Fig. 21 Waveforms of speed-controlled SMC SRM drive with load change

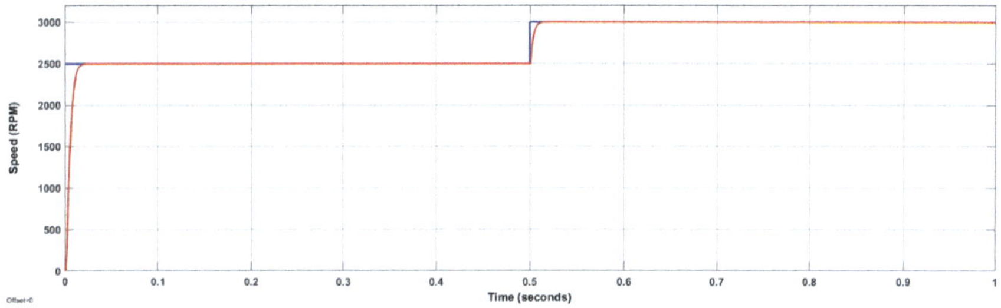

Fig. 22 Speed waveform of speed-controlled SMC SRM drive with sudden change in reference speed

Transient response with change in reference speed:

Transient response of the speed-controlled SMC drive with a sudden change in reference speed value from 2500 to 3000 RPM at 0.5 s is depicted in Fig. 22 with a load torque of 3 Nm. It is very well clear that the speed change is tracked in 15 ms time. The peak current increases from 9.5 A to 20 A and there is an increase in torque ripple from 3.25 Nm to 5 Nm.

8 Conclusions and Inferences

Mathematical modelling of SRM and small signal model of machine are presented in this chapter. The simulation results of different stages of implementation of the speed control loop in MATLAB Simulink are presented. Open loop control of machine-driven with an asymmetric converter is implemented where the gate signals are generated dependent on rotor position feedback signals. The current control inner loop of SRM is implemented with a hysteresis current controller and sliding-mode current controller is simulated.

Table 2 Transient response analysis

System	Input variation			Speed change (RPM)		Speed settling time (s)
	Parameter	Initial	Final	Initial	Final	
Open loop	T_L *(Nm)*	0	3	9360	8300	1.5
Current controlled HCC drive	T_L *(Nm)*	3	5	6425	5545	3
	I_{ref} *(A)*	20	30	6425	7336	2
Current controlled SMC drive	T_L *(Nm)*	3	5	6338	5517	2.6
	I_{ref} *(A)*	20	30	6338	7334	1.8
Speed controlled HCC drive	T_L *(Nm)*	3	10	2500	2500	2
	ω_{ref} *(RPM)*	2500	3000	2500	3000	0.023
Speed controlled SMC drive	T_L *(Nm)*	3	10	2500	2500	1
	ω_{ref} *(RPM)*	2500	3000	2500	3000	0.015

The results of the transients to variation in various input conditions are summarized in Table 2 and transients due to load and reference current are studied. For the current control inner loop, there is no control over the speed of SRM. The speed control is thereby implemented with both the current controllers, along with a PID outer loop speed control and transients due to the change in load and current are observed. The parameters T_L is load torque, I_{ref} is reference current and ω_{ref} is speed reference value. The settling time for speed is reduced using sliding mode controller compared to hysteresis current controller.

Currently, SRMs are attracting attention mainly due to their magnet-free design in the research community, lower-cost, high-efficiency and high-speed applications. SRM has the longest constant power speed range and is robust at high speeds and high temperature conditions. The high torque-ripple may restrict its use in EVs. In this chapter, PID speed controller and two different current controllers are simulated and presented. Second order sliding mode current controller reduces the current ripple leading to a reduction in torque ripple. Further torque control techniques can be incorporated with the system to enable constant torque operation. Different nonlinear control strategies can help to further reduce the torque-ripple and develop an efficient and cost-effective SRM drive for EVs. New research in AI-based techniques for control can be found in [21, 22].

References

1. Mukhopadhyay J, Choudhuri S, Sengupta S (2015) Drive strategies for switched reluctance motor—a review. In: Michael Faraday IET international summit, pp 132–135. https://doi.org/10.1049/cp.2015.1619
2. Velmurugan G, Bozhko S, Yang T (2018) A review of torque ripple minimization techniques in switched reluctance machine. In: IEEE international conference on electrical systems for

aircraft, railway, ship propulsion and road vehicles, international transportation electrification conference, Nottingham, UK, pp 1–6. https://doi.org/10.1109/ESARS-ITEC.2018.8607614
3. Husain I (2002) Minimization of torque ripple in SRM drives. IEEE Trans Ind Electron 49(1):28–39
4. Valencia DF, Tarvirdilu-Asl R, Garcia C, Rodriguez J, Emadi A (2021) Vision, challenges, and future trends of model predictive control in switched reluctance motor drives. IEEE Access 9:69926–69937. https://doi.org/10.1109/ACCESS.2021.3078366
5. Ahn JW, Lukman GF (2018) Switched reluctance motor: research trends and overview. CES Trans Electr Mach Syst 2(4):339–347. https://doi.org/10.30941/CES-TEMS.2018.00043
6. Husain I (2018) Electric and hybrid vehicles design fundamentals. CRC Press, Boca Raton
7. Cunningham JD (2011) Switched reluctance motor drive circuit evaluation criteria for vehicle efficiency and responsiveness. MSE thesis, University of Texas, Austin
8. Ahn JW (2001) Torque control. InTech publisher (Chapter 8)
9. Suresh S, Peter J, Chellappan V (2022) Mathematical modeling and speed control of 6/4 switched reluctance motor drive. In: International conference on futuristic technologies in control systems and renewable energy, pp 1–6. https://doi.org/10.1109/ICFCR54831.2022.9893644
10. Krishnan R (2001) Switched reluctance motor drives. CRC Press, Boca Raton
11. Narayanan A, Athulya PC, Rahmath TA, Vismaya A, Kurian S, Chellappan V (2022) SRM motor control using conventional and full-order-terminal SMC controller. In: 2nd Asian conference on innovation in technology, pp 1–7. https://doi.org/10.1109/ASIANCON55314.2022.9909124
12. Kurian S, Nisha GK (2015) Torque ripple minimization of SRM using torque sharing function and hysteresis current controller. In: International conference on control communication and computing, pp 149–154. https://doi.org/10.1109/ICCC.2015.7432884.
13. Kurian S, Nisha GK (2022) Current control with hysteresis controller and sliding mode controller in SRM. In: IEEE international conference on signal processing, informatics, communication and energy systems, pp 1–6. https://doi.org/10.1109/SPICES52834.2022.9774155
14. Nashed MNF, Mahmoud SM, El-Sherif SM, Abdel-Aliem ES (2014) Hysteresis current control of switched reluctance motor in aircraft applications. Int J Power Electron Drive Syst 4(3):376–392
15. Levant A (1997) Higher order sliding: collection of design tools. In: European control conference, Brussel, Belgium
16. Edwards C, Spurgeon SK (1998) Sliding mode control: theory and applications. CRC Press
17. Rain X, Hilairet M, Talj R (2010) Second order sliding mode current controller for the switched reluctance machine. In: IEEE conference of industrial electronics society, pp 3301–3306. https://doi.org/10.1109/IECON.2010.5675042
18. Levant A (1993) Sliding order and sliding accuracy in sliding mode control. Int J Control 58:12471263
19. Moreno JA, Osorio M (2008) A Lyapunov approach to second-order sliding mode controllers and observers. In: 47th IEEE conference on decision and control, pp 2856–2861. https://doi.org/10.1109/CDC.2008.4739356
20. Kurian S, Nisha GK (2015) Closed loop operation of switched reluctance motor with hysteresis controller. Int J Appl Eng Res 10(17):37645–37650
21. Bhausaheb RM (2021) Speed control of SRM for hybrid electric vehicle using artificial intelligence. In: 12th International conference on computing communication and networking technologies, Kharagpur, India, pp 01–05. https://doi.org/10.1109/ICCCNT51525.2021.9579857
22. Kandhasamy S (2020) Machine learning based SRM control using FPGAs for torque ripple minimization. In: International conference on artificial intelligence in information and communication, pp 675–680. https://doi.org/10.1109/ICAIIC48513.2020.9065241

Robotic Arm Inverse Kinematics Through Graphical Approach

G. Madan, A. Venkata Vishnu, J. Kumar Chandra, R. Ashutosh, and Y. V. S. V. Puneeth

Abstract Automation and Robotics have a lot of scope in the present scenario; there are many fields of applications in the industrial sector for robot manipulators like material handling, space manipulation, welding, grinding, spray painting, vision inspection, etc. To achieve the above-stated goal, one should have sound knowledge of inverse kinematics to control a robot mechanism. This chapter deals with the forward and inverse kinematics of a 4 degree of freedom spatial robotic arm. To determine the end effector position, the D-H frameworks are used for forward kinematics, and a graphical approach was used for inverse kinematics solution. The graphical approach is obtained by solving the values of joint variables at a desired end effector position using trigonometry. The movement of the robotic arm is actuated with stepper motors by using the microcontroller and Raspberry pi.

Keywords Automation and robotics · DH parameters · Forward kinematics · Inverse kinematics · Graphical approach · Raspberry Pi

G. Madan (✉) · A. V. Vishnu · J. K. Chandra · R. Ashutosh · Y. V. S. V. Puneeth
DigiTele Networks Pvt. Ltd, 1071, Rd Number 44, CBI Colony, Jubilee Hills, Hyderabad, Telangana, India
e-mail: madan.g@digitelenetworks.com

A. V. Vishnu
e-mail: venkata.a@digitelenetworks.com

J. K. Chandra
e-mail: kumarchandra.j@digitelenetworks.com

R. Ashutosh
e-mail: ashutosh.r@digitelenetworks.com

Y. V. S. V. Puneeth
e-mail: puneeth.y@digitelenetworks.com

A. Kumar et al. (eds.), *Advances in Mechatronics Systems*, Emerging Trends in Mechatronics, https://doi.org/10.1007/978-981-97-8472-1_11

1 Introduction

In the present scenario, the usage of robots in engineering, medical, service sectors, etc. has increased remarkably, as it enables flexibility in terms of high quality at low cost and time [1, 2]. The demand for assistive robots is rising as they enable individuals to work independently and accomplish tasks that would otherwise be impossible [3]. A robot arm consists of a flexible series of links that are associated through joints, it consists of an end-effector that moves in space freely and the other end fixed to the ground, the movement of the arm may be revolute and prismatic. Robotic arms are used in numerous areas and have the ability to perform tasks with high precision and speed. Although, robotic arm controlling is difficult, which path and position planning of the robot arm plays a crucial role [4–7]. Kinematic analysis is the direct correlation between end-effector space and joint space which plays a crucial role in the control and motion planning of the arm [8, 9]. Displacement Kinematics consists of forward and inverse kinematics [6].

Several investigations have been carried out in order to study the kinematic model, Table 1 lists out the outcomes of the literature and it is observed that Inverse kinematics plays an important part in determining precise movement and time calculation; Computing an inverse kinematic solution is a challenging and complicated task [16]. Inverse kinematics can be solved using different methodologies like iterative, geometric, algebraic, etc. [17]. In the present work, a robotic arm was developed in order to study the forward and inverse kinematics of a four degree of freedom arm and was derived by making use of a graphical method.

2 Methodology

To derive a fully graphical solution, the investigation embraces a four-degree-of-freedom robotic arm, consisting of four revolute links including the end effector. The construction of the arm is shown in Fig. 1. Most of the commercial robotic arms are of the same construction, used in many industrial applications as shown in Fig. 1. In order to achieve the set of possible solutions, Forward kinematics is carried out utilizing the Denavit-Hartenberg (D-H) method and inverse kinematics using the graphical method.

2.1 *Forward Kinematics*

Denavit-Hartenberg (D-H) process is used to determine the spatial correlation with respect to two relative joints, the link notation plays a crucial role and when a frame is in contact it consists of two joints at both ends which have an association with respect to position and geometry of the joints [13]. Considering the Denavit-Hartenberg

Table 1 Outline of literature survey

Authors	Outcomes
Abaas et al. [1] and Fusic et al. [10]	• Forward, inverse, and velocity kinematics analysis of a 5-DOF robotic arm are developed • The Denavit-Hartenberg (DH) parameters are used to determine the forward kinematics, while an algebraic solution is applied for the inverse kinematics to find the position and orientation of the end effector • The Jacobian matrix is utilized to calculate the velocity kinematics of the robotic arm • The position and orientation of the end effector are calculated using MATLAB software based on the DH parameters • The results indicate that the shoulder joint has the most significant effect on the velocity of the robotic arm compared to other joints, with the maximum error in the position of the end effector occurring along the z-axis and the minimum error along the y-axis
Myint et al. [2] and Mahakalkar et al. [11]	• The importance of kinematic methods are studied with a pick and place robotic arm • Concluded that Kinematic methods provide an analytical description of the manipulator's geometric motion relative to a fixed robot coordinate system, without considering the forces or moments causing the movements • The position control of the pick-and-place robotic arm has been fully tested, yielding precise results
Barua et al. [12]	• A 3DOF robotic arm is designed and analyzed using DH parameters and a kinematics model, with its movements simulated using software • The arm is programmable with the application of sensors that measure the distance between the robotic arm and the object, providing feedback to the microcontroller • Using this feedback, the microcontroller determines the motor movements needed to efficiently pick and place the object
Ting et al. [13]	• A kinematic model was developed to accurately calculate the desired position of the end-effector based on 4 DOF parameters • The performance of the developed kinematic model is evaluated using a simulated robot arm, analyzing whether errors fall within acceptable ranges • The study successfully models the forward and inverse kinematic solutions for a 4-DoF articulated robot arm, laying the theoretical groundwork for future analysis and research in this field
Roshanianfard et al. [14] and Qureshi et al. [15]	• The dynamic aspects of the system were analyzed using standard mechanical formulas, focusing on parts requiring torque calculations and mass center coordination through conventional machine design methods • Denavit-Hartenberg methodology was employed for both forward and inverse kinematics calculations • To optimize torque requirements, components were designed using various materials and mass center adjustments, with performance comparisons conducted • Feedback from an inverse kinematics algorithm confirmed optimal operational performance of the robotic arm

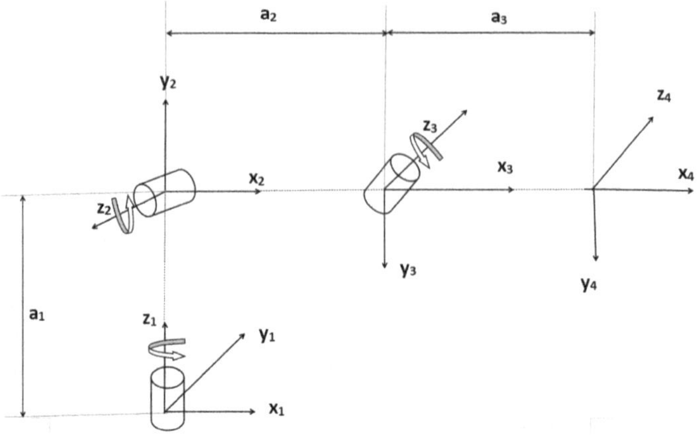

Fig. 1 Kinematic diagram for robo arm

Table 2 D-H paramaters

Number of joints	θ	α	r	d
1	θ_1	90	0	17
2	θ_2	0	30	0
3	θ_3	0	25	0

parameters, a fixed base frame is developed and frames are joined accordingly through its end joints as shown in Fig. 1 and its parameters are listed in Table 2. To compute X, Y, and Z axes Homogeneous transformation matrix is developed based on the D-H parameters, it computes the transformation of frame i respective to the frame i−1 shown in Eq. 1 [3]. Where 'c' represents cos and 's' sin respectively [9].

$$T_i^{i-1} = \begin{bmatrix} c\ (\theta_i) & -s(\theta_i)c(\alpha_i) & s(\theta_i)s(\alpha_i) & \alpha_i c(\theta_i) \\ s\ (\theta_i) & c(\theta_i)c(\alpha_i) & -c(\theta_i)s(\alpha_i) & \alpha_i s(\theta_i) \\ 0 & s(\alpha_i) & c(\alpha_i) & d_i \\ 0 & 0 & 0 & 1 \end{bmatrix} \tag{1}$$

2.2 *Inverse Kinematics*

The arm is of 3D spatial relationship which consists of end effector position X, Y & Z. The solution is divided into two cases. Figure 2 shows the first case of the calculation, to get the angle (θ_1) which can be generated through the top view of the kinetic diagram from Fig. 1. Here r is the vector which denotes the end effector

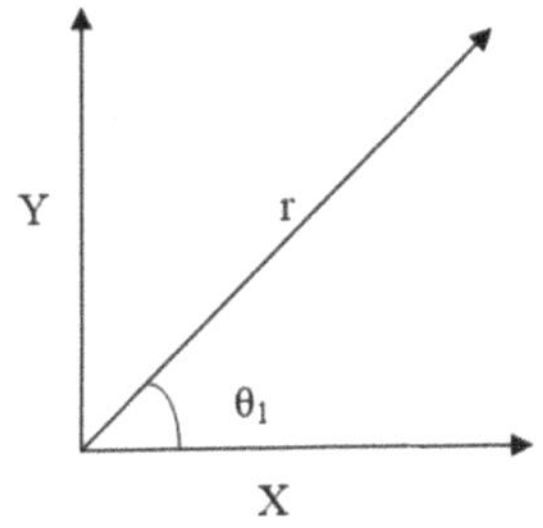

Fig. 2 Top view of kinetic diagram

position. The X and Y plane movement of the robot arm rests on the angle (θ_1) for joint 1, its surface projection for arm movement, the angle, θ_1 can be calculated as follows,

$$\theta_1 = \tan^{-1}\frac{y}{x} \tag{2}$$

Here θ_1 is the angle of rotation of the arm base; ranges between $-150°$ and $150°$. From Fig. 2 we can also write as

$$\mathrm{r} = \sqrt{X^2 + Y^2} \tag{3}$$

The next case pertains to calculating the angles for joints a_2 and a_3 with corresponding angles θ_2 and θ_3. For obtaining the solution for θ_2 and θ_3, the front view of the kinematic diagram is considered as shown in Fig. 3. Axis 'r' is the new axis developed by merging X and Y-axis.

$$Z_1 = Z - a_1 \tag{4}$$

$$\theta_2 = \emptyset_1 - \emptyset_2 \tag{5}$$

$$\emptyset_1 = \tan^{-1}\left(\frac{Z1}{r}\right) \tag{6}$$

$$\emptyset_2 = \mathrm{a}\tan2\left(\frac{\sin\emptyset_2}{\cos\emptyset_2}\right) \tag{7}$$

$$\sin\emptyset_2 = \sqrt{1 - \cos^2\emptyset_2} \tag{8}$$

From cosine rule

$$\cos\emptyset_2 = \frac{a_3^2 - a_2^2 - r_1^2}{-2a_2r_1} \tag{9}$$

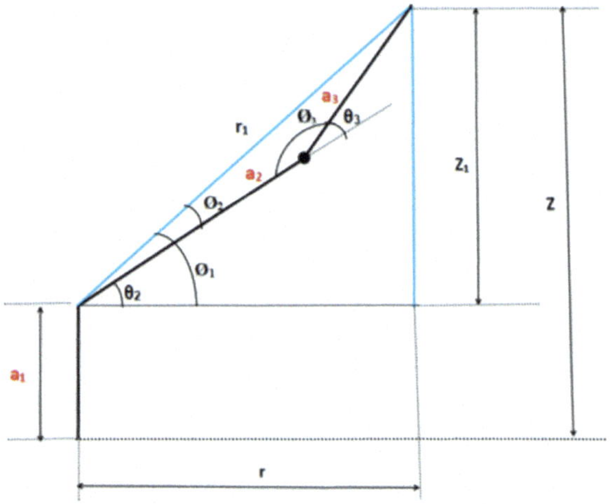

Fig. 3 Front view of kinetic diagram

where, $r_1 = \sqrt{Z_1^2 + r^2}$

From Fig. 3, we can also write

$$\theta_3 = 180^\circ - \varnothing_3 \tag{10}$$

$$\varnothing_3 = \operatorname{a tan2}\left(\frac{\sin\varnothing_3}{\cos\varnothing_3}\right)$$

$$\sin \varnothing_3 = \sqrt{1 - \cos^2\varnothing_3} \tag{11}$$

From cosine rule

$$\cos \varnothing_3 = \frac{r_1^2 - a_2^2 - a_3^2}{-2a_2a_3} \tag{12}$$

The two possible configurations "elbow up" and "elbow down" for the robot arm are considered in the calculation of θ_2 and θ_3.On solving the equations from 2 to 12 the 3 angles will be identified. The angle range for θ_2 is 0° to 120° and θ_3 is between − 100° to 100°, respectively.

3 Assembling and Controlling of Robotic Arm

The mechanical structure of the robo arm is developed with the help of wood pieces. Different types of electric motors are considered based on the robo arm physical and mechanical characteristics to create a movement. The required torque of a motor is calculated for each stepper motor were two stepper motors are used with a capacity of

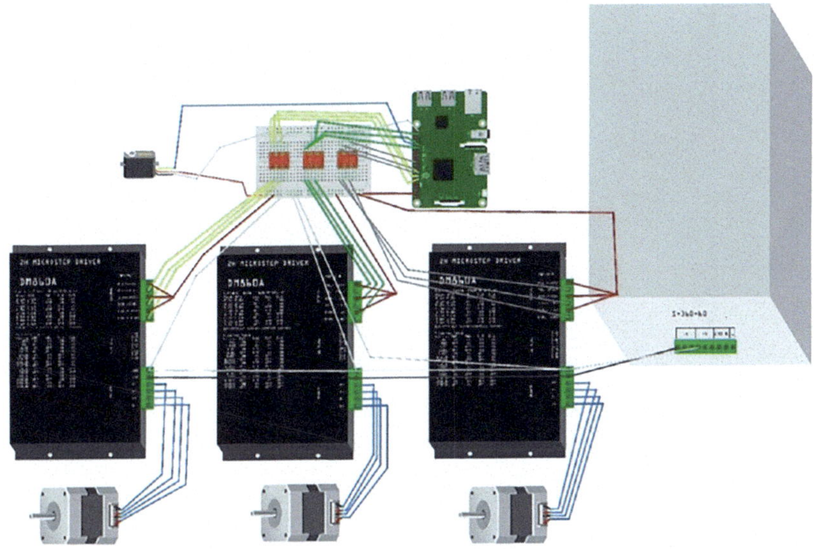

Fig. 4 Circuit diagram of raspberry pi with motors, driver and power supply

60 kg-cm, one stepper of 45 kg-cm, and one servo motors for end effector. Metal shaft couplings are used in order to connect the wood to stepper motor shafts. The Micro step drivers DM 860B maintain the current and control the three stepper motors to make it work effectively. The Servo motor is connected to a raspberry pi and the detailed circuit diagram is shown in Fig. 4. Here the link lengths a_1, a_2 and a_3 are 17 cm, 30 cm, and 25 cm respectively.

A continuous power supply for stepper is provided with the help of RMCS-4024 24 V 10A Industrial Power Supply as shown in Fig. 4. The Microprocessor used is Raspberry Pi 4 Model B.

A Python code is developed based on the forward kinematics shown in Fig. 5 and inverse kinematics from Fig. 6 calculations and processed in Raspberry Pi 4 to control the Robo arm.

4 Experimental Results

The robotic arm is developed with three stepper motors, Raspberry Pi, Power supply, drivers, and one servo which is the end effector with an 360 degrees rotation angle, which can be used in many applications as shown in Fig. 7 and a lot of experiments for both forward and inverse kinematics have been conducted by varying different end effector positions and different joint angles. From Tables 3 and 4, the connection linking the forward and inverse kinematic model developed is known to be valid. The position results of the robotic arm end effector were tabulated in Table 5 which

```
import numpy as np
import math

def check_the_dist(theta_1, theta_2, theta_3):
    a1 = 17  # Length of link 1
    a2 = 30  # Length of link 2
    a3 = 25  # Length of link 3

    d_h_table = np.array([[np.deg2rad(theta_1), np.deg2rad(90), 0,
17],[np.deg2rad(theta_2),np.deg2rad(0), 30, 0],[np.deg2rad(theta_3), np.deg2rad(0),
25, 0]])
    # Homogeneous transformation matrix from frame 0 to frame 1
    i = 0
    homgen_0_1 = np.array([[np.cos(d_h_table[i, 0]), -np.sin(d_h_table[i, 0]) *
np.cos(d_h_table[i, 1]),np.sin(d_h_table[i, 0]) * np.sin(d_h_table[i, 1]),
d_h_table[i, 2] * np.cos(d_h_table[i, 0])],[np.sin(d_h_table[i, 0]),
np.cos(d_h_table[i, 0]) * np.cos(d_h_table[i, 1]),-np.cos(d_h_table[i, 0]) *
np.sin(d_h_table[i, 1]), d_h_table[i, 2] * np.sin(d_h_table[i, 0])],[0,
np.sin(d_h_table[i, 1]), np.cos(d_h_table[i, 1]), d_h_table[i, 3]],[0, 0, 0, 1]])

    # Homogeneous transformation matrix from frame 1 to frame 2
    i = 1
    homgen_1_2 = np.array([[np.cos(d_h_table[i, 0]), -np.sin(d_h_table[i, 0]) *
np.cos(d_h_table[i, 1]),np.sin(d_h_table[i, 0]) * np.sin(d_h_table[i, 1]),
d_h_table[i, 2] * np.cos(d_h_table[i, 0])],[np.sin(d_h_table[i, 0]),
np.cos(d_h_table[i, 0]) * np.cos(d_h_table[i, 1]), -np.cos(d_h_table[i, 0]) *
np.sin(d_h_table[i, 1]), d_h_table[i, 2] * np.sin(d_h_table[i, 0])],[0,
np.sin(d_h_table[i, 1]), np.cos(d_h_table[i, 1]), d_h_table[i, 3]],[0, 0, 0, 1]])

    # # Homogeneous transformation matrix from frame 2 to frame 3
    i = 2
    homgen_2_3 = np.array([[np.cos(d_h_table[i, 0]), -np.sin(d_h_table[i, 0]) *
np.cos(d_h_table[i, 1]),np.sin(d_h_table[i, 0]) * np.sin(d_h_table[i, 1]),
d_h_table[i, 2] * np.cos(d_h_table[i, 0])],[np.sin(d_h_table[i, 0]),
np.cos(d_h_table[i, 0]) * np.cos(d_h_table[i, 1]), -np.cos(d_h_table[i, 0]) *
np.sin(d_h_table[i, 1]), d_h_table[i, 2] * np.sin(d_h_table[i, 0])],[0,
np.sin(d_h_table[i, 1]), np.cos(d_h_table[i, 1]), d_h_table[i, 3]],[0, 0, 0, 1]])
    homgen_0_3 = homgen_0_1 @ homgen_1_2 @ homgen_2_3
    print(homgen_0_3)
    x = homgen_0_3[0][-1]
    y = homgen_0_3[1][-1]
    z = homgen_0_3[2][-1]
    return x,y,z

print(x,y,z)
```

Fig. 5 Forward kinematics Python code

shows good results in determining the position with respect to arm movement and the equations derived theoretically.

```
import numpy as np
import math
def get_angles(x, y, z):
    a1 = 17  # Length of link 1
    a2 = 30  # Length of link 2
    a3 = 25  # Length of link 3

    theta_1 = math.atan2(float(y), float(x))
    print(math.degrees(theta_1))
    z1 = float(z) - float(a1)
    r = math.sqrt(float(x) ** 2 + float(y) ** 2)
    r1 = math.sqrt(float(z1) ** 2 + float(r) ** 2)
    cphy2 = float((float(a3) ** 2 - float(a2) ** 2 - float(r1) ** 2) / (-2 * float(a2)
* float(r1)))
    sphy2 = float(math.sqrt(1 - cphy2 ** 2))
    phy1 = float(math.atan2(float(z1), float(r)))
    phy2 = float(math.atan2(sphy2, cphy2))
    theta_2 = phy1 - phy2
    print(math.degrees(theta_2))
    cphy3 = float((float(r1) ** 2 - float(a3) ** 2 - float(a2) ** 2) / (-2 * float(a2)
* float(a3)))
    sphy3 = float(math.sqrt(1 - cphy3 ** 2))
    phy3 = float(math.atan2(sphy3, cphy3))
    theta_3 = float(math.pi - phy3)
    print(math.degrees(theta_3))
    return math.degrees(theta_1), math.degrees(theta_2), math.degrees(theta_3)
```

Fig. 6 Inverse kinematics Python code

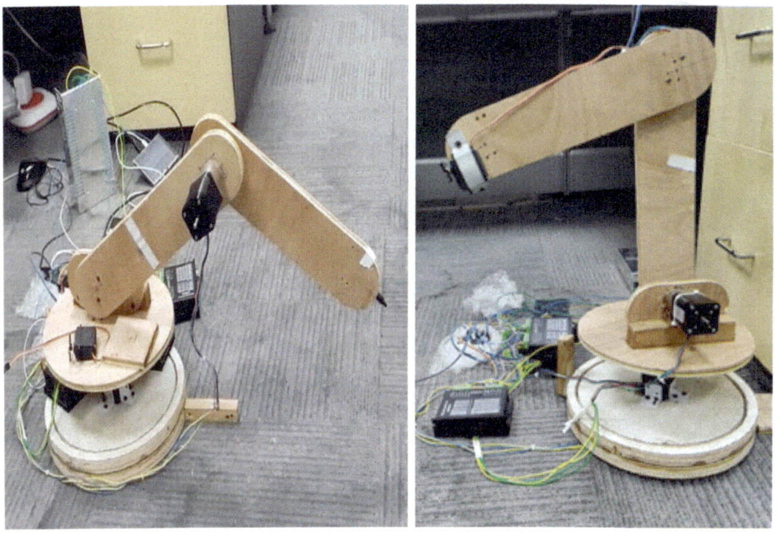

Fig. 7 Working model of robo arm

Table 3 Results of forward kinematic model

End effectors positions (cm)			Angles (°)		
X	Y	Z	θ_1	θ_2	θ_3
55	0	17	0	0	0
42	27	29	30.25	27.24	51.34
34	33	35	32.9	−8.6	72.9
−15	32	22	115.1	−35.6	99.6
−22	16	38	143.9	−7.41	103.25

Table 4 Results of inverse kinematic model

End effectors positions (cm)			Angles (°)		
θ_1	θ_2	θ_3	X	Y	Z
0	0	0	55	0	17
30.25	27.24	51.34	42	26.97	29.08
32.9	−8.6	72.9	34.08	−22	35.04
115.1	−35.6	99.6	−14.9	32.01	22.06
143.9	−7.41	103.25	−21.98	16.02	38

Table 5 Position results of end effector

	Angles (°)			End effector positions of robo arm (cm)			End effector position (theoretically)		
Case	θ_1	θ_2	θ_3	X	Y	Z	X	Y	Z
1	0	0	0	55	0	17	55	0	17
2	30.25	27.24	51.34	42	26.97	29.08	42	26.7	29
3	32.9	−8.6	72.9	34.08	−22	35.04	34	−22	35
4	115.1	−35.6	99.6	−14.9	32.01	22.06	−14	32	22
5	143.9	−7.41	103.25	−21.98	16.02	38	−21	16	38

5 Conclusion

This current work is successfully analyzed for both forward and inverse kinematic models of robotic arms using a graphical approach. Correlation linking the forward and inverse kinematic models was analyzed and verified with respect to the end effector position. The initiated graphical method is facile to acknowledge compared with other approaches and now days apart from the above approach researchers are looking towards ROS in order to control the rover framework.

6 Future Scope

In the current work using a kinematic model the movement of a robotic arm is obtained, apart from the above there are a lot of potential advancements and scope in robotic arm technology, such as improved materials, actuators, sensors, and control algorithms which would enhance the arm's performance, efficiency and reliability. Integrating artificial intelligence and machine learning could enable robotic arms to optimize tasks and adapt to dynamic environments more effectively. As industry is looking towards sustainability aspects like energy efficiency, waste reduction, and mitigating environmental impact which needs to be considered when developing a robotic arm. Exploring various strategies aimed at reducing the cost of robotic arm systems to enhance accessibility for smaller businesses and industries.

References

1. Abaas TF, Khleif AA, Abbood MQ (2020) Kinematics analysis of 5 DOF robotic arm. Eng Technol J 38(03/Part A):412–422
2. Myint KM, Htun ZMM, Tun HM (2016) Position control method for pick and place robot arm for object sorting system. Int J Sci Technol Res 5(6):57–61
3. Kruthika K, Kumar BK, Lakshminarayanan S (2016) Design and development of robotic arm. Published on International Conference on Circuits, Controls, Communication and Computing (I4C). https://doi.org/10.1109/CIMCA.2016.8053274
4. An HS, Lee JH, Lee C, Seo TW, Lee JW (2017) Geometrical kinematic solution of serial spatial manipulators using screw theory. Mech Mach Theory 116:404–418
5. Lopez-Franco C, Hernandez-Barragan J, Alanis AY, Arana-Daniel N, Lopez-Franco M (2018) Inverse kinematics of mobile manipulators based on differential evolution. Int J Adv Robot Syst:1–22
6. Lopez-Franco C, Hernandez-Barragan J, Alanis AY, Nancy A (2018) A soft computing approach for inverse kinematics of robot manipulators. Eng Appl Artif Intell 74:104–120
7. Zhang H, Xia Q, Sun J, Zhao Q (2022) A fully geometric approach for inverse kinematics of a six degree-of-freedom robot arm. ICAMMT-2022 J Phys: Conf Ser 2338:012089, IOP Publishing. https://doi.org/10.1088/1742-6596/2338/1/012089
8. Funda J, Taylor R, Paul R (1990) On homogeneous transforms, quaternions, and computational efficiency. IEEE Trans Robot Autom 6:382–388. https://doi.org/10.1109/70.56658
9. Singh S, Singla E (2016) Service arms with unconventional robotic parameters for intricate workstations: optimal number and dimensional synthesis. J Robot:1–11
10. Fusic SJ, Giri J, Makki E, Sitharthan R, Murugesan S, Bhowmik A (2024) Momordica charantia leaf disease detection and treatment using agricultural mobile robot. AIP Adv 14(4). https://doi.org/10.1063/5.0190928
11. Mahakalkar SG, Giri JP, Sunheriya N, Lande P, Chdge R, Mahatme C, Sathish T (2024) Propulsion effectiveness using propeller selection, ducted propellers and tilting mechanism in a quadrotor drone. Lecture notes in mechanical engineering, pp 399–410. https://doi.org/10.1007/978-981-97-3173-2_26
12. Barua R, Kumar SS, Mallik A, Singh A (2021) Experimental analysis the dynamic model of 3DOF robotic arm. J Mech Robot 6(1):20–26, e-ISSN: 2582–2187
13. Ting HZ, Hairi M, Zaman M, Ibrahim MF, Moubark AM (2021) Kinematic analysis for trajectory planning of open-source 4-DoF robot arm. (IJACSA) Int J Adv Comput Sci Appl 12(6)

14. Roshanianfard A, Noguch N (2018) Kinematics analysis and simulation of a 5DOF articulated robotic arm applied to heavy products harvesting. J Agric Sci 24:91–104
15. Qureshi MJ, Kutty AV, Giri J, Dapke A, Giri P, Chadge RB, Sunheriya N (2023) Public acceptance of drones: an approach. CRC Press, pp 493–503. https://doi.org/10.1201/9781003358596-55
16. KA GJ, Melingui A, Lakhal O, Kom M, Merzouki R (2020) A learning framework to inverse kinematics of redundant manipulators. IFAC-Papers online 53(2):9912–9917
17. Angela Sodemann AA (2020) How to assign Denavit-Hartenberg frames to robotic arms. https://automaticaddison.com/how-to-assign-denavit-hartenberg-frames-to-robotic-arms/. Accessed 2022